From California's Gold Fields to the Mendocino Coast

From California's Gold Fields to the Mendocino Coast

A SETTLEMENT HISTORY ACROSS TIME AND PLACE

Samuel M. Otterstrom

UNIVERSITY OF NEVADA PRESS *Reno & Las Vegas*

University of Nevada Press | Reno, Nevada 89557 USA
www.unpress.nevada.edu

Cover design by Rebecca Lown

LIBRARY OF CONGRESS CATALOGING-IN-PUBLICATION DATA
Names: Otterstrom, Samuel, author.
Title: From California's gold fields to the Mendocino Coast : a settlement history across time and place / Samuel M. Otterstrom.
Description: First edition. | Reno : University of Nevada Press, [2017] | Includes bibliographical references and index.
Identifiers: LCCN 2016054335 (print) | LCCN 2016055116 (e-book) | ISBN 978-1-943859-28-3 (cloth : alk. paper) | ISBN 978-0-87417-469-4 (e-book)
Subjects: LCSH: California, Northern—Discovery and exploration. | California, Northern—History—19th century. | Frontier and pioneer life—California, Northern. | Land settlement—California, Northern—History—19th century.
Classification: LCC F867.5 .O78 2017 (print) | LCC F867.5 (ebook) | DDC 979.4—dc23
LC record available at https://lccn.loc.gov/2016054335

The paper used in this book meets the requirements of American National Standard for Information Sciences— Permanence of Paper for Printed Library Materials, ANSI/NISO Z39.48-1992 (R2002).

FIRST PRINTING

Manufactured in the United States of America

Contents

Illustrations

TABLES

From California's Gold Fields
to the Mendocino Coast

1

The Scales of California Settlement History

New Hope is an optimistic place name, and hope and optimism likely pervaded the thoughts of the approximately 230 Mormons who sailed with Samuel Brannan from New York to California on the ship *Brooklyn*, arriving in Yerba Buena (San Francisco) July 31, 1846. About twelve families moved on to the Stanislaus River to establish a new community in anticipation that many Mormons would follow them there (J. Davies 1984, 8–10). Despite its name, New Hope Colony did not last very long. The bulk of the Mormons never came to settle in California but instead stopped in the valley of the Great Salt Lake.

By the time the announcement of gold in the Sierra Nevada had become public in 1848, New Hope was gone. Its demise was assured by the colony's poor location (plagued by mosquitoes and malaria), conflict with Sam Brannan, and the news that the main body of Mormons was not coming to California (Owens 2004, 47–48). In contrast, the Mormons who remained in Yerba Buena found more economic success as the peninsula was poised to grow, due to the event soon to occur in the Sierra foothills to the east (Stellman 1953, 64–65).

The great "new hope" for the future growth of the newly U.S. California began at an obscure place, Sutter's Mill, that did not fade from the map as New Hope Colony had done. Incidentally, it too had a Mormon presence. John Sutter, needing laborers to build his sawmill on the banks of the American River, hired a number of Mormons (formerly of the Mormon Battalion) who had traveled together to California to fight in the Mexican-American war. They fought no battles, because the war had already ended by the time they arrived in San Diego on January 29, 1847 (J. Davies 1984, 6). Many of these Mormons decided to stay in California for a time to work, and at least five hired on at Sutter's Mill. They were on the job when James Marshall discovered gold in the tailrace at the mill site in January of 1848, and two of them recorded some of the earliest accounts of the discovery in their personal writings.

Regardless, instead of a new hope emanating from an agricultural colony in the San Joaquin valley, it sounded with a thunderous clap from the small sawmill at Coloma, recorded and expanded along the river by Mormon laborers. In San Francisco, Samuel Brannan promoted and advertised the discovery with verve, for he stood ready to make his transitory millions through the newspaper business and by selling supplies to the miners who soon would come. Brannan's mercurial rise and fall is one of the great stories of early California history.

But for every Sutter, Marshall, and Brannan, there were thousands of relatively unknown characters who also made their mark on the promising California landscape. Most of their movements are lost to history, but dependence and interaction between the mines in the Sierra Nevada foothills and the San Francisco Bay area were ongoing and significant. One Mary Ann Fisher Cheney, a Mormon immigrant who arrived on the *Brooklyn*, wrote in 1850 of how she and her husband panned for gold in the Sierra and raised produce on their farm at San Jose. She noted that over the winter of 1849–1850, they were able to sell a dozen eggs for the incredibly high price of $6. These exorbitant prices did not last long, as farming was suddenly a very attractive option for the forty-niners and other immigrants (Owens 2004, 253–54). Mary Cheney and her husband are just two examples of the thousands of people who arrived to the valleys blossom with produce and livestock, to provoke the rivers and streams to give forth their riches, or both. Thus, the symbiotic cycle between the mountains and valleys of California was established. More importantly, it was often the lot of successful or discouraged miners alike to be integrally connected with the economic development and spatial dependency that ensued because of the monetary courses they took.

PEOPLING OF CALIFORNIA ACROSS TIME AND SPACE

This region of California certainly has a story to tell. The story is hidden beneath a wave of untold numbers of gold miners, merchants, farmers, politicians, carpenters, and ever more people from various backgrounds and corners of the world. Amidst this mass of historical data is an intricately woven tapestry of interrelated people and events that literally created this dynamic state. At its most basic level, northern California can be geohistorically dissected into composite elements of individual persons acting in their defined portions of the landscape at a particular point of time. Very quickly, study at this individual level requires an agglomeration of individuals, first in family units, next in small communities or neighborhoods, followed by counties or cities, and finally a regional system. Just as in the human body, where no one atom or cell can act individually without affecting its surrounding elements, the history of nations has been written and shaped both by the most incongruous farmer and the exceptionally boisterous politician.

This book explores the lives of the many people who made the valleys of California bloom within a wide variety of human landscapes. Dozens of excellent books have been written concerning the early growth and development of the state, from sweeping comprehensive histories (Bean and Rawls 1983; Brands 2002; Lavender 1972; Roske 1968; Starr 1973) to regionalized and geographically focused monographs (Brechin 1999; Fradkin 1995; Hardwick and Holtgrieve 1996; Holliday 1999; Lotchin 1974). This book adds a particular perspective to that large and competent body of literature. It illustrates how the new settlers' initial land-altering presence persisted over time through their family relationships, and how the early pioneers and miners periodically transformed themselves in response to constantly shifting economic paradigms. By multiplying these individualistic experiences across the far-flung reaches of the California system into larger and larger scales, this book uncovers the secrets that propelled a sleepy post–Mexican outpost into a flurry of restless, youthful growth, which has continued virtually unchecked until today.

As the unending throng of people pushed into California beginning in late 1848 and early 1849 and throngs surging into the next several years, they changed the region forever. Their impact on the rivers and forested slopes of the Sierra Nevada mountain foothills was significant, but most of the players in this drama were like the butterflies that cover the hillsides in the fairer months but disappear in the winter. The miners were constantly on the lookout for better diggings in other sparkling streams in other valleys, and so permanence of address was not common. This seething, never-resting mass of opportunistic argonauts often made its most lasting impacts in the cities and farms of other parts of the state after individual luck ran out, or when people became tired of the uncomfortable life of a miner. San Francisco and Sacramento would likely have not grown so spectacularly were it not for the immigrants and the businesses associated with mining. In this way, all of northern California was intertwined and interrelated in the nearly living regional organism that matured into an economically innovative and increasingly dynamic spatial system (fig. 1.1).

It is hard to fathom the movement of thousands of people across the face of America in search of "easy" riches. Money has the ability to motivate people, where other incentives do not. Try to imagine the sheer magnitude of the multitudes of men and women who streamed across the continent on foot or horse.

FIGURE 1.1. Northern California. Ryan Shields, ThinkSpatial, BYU Geography.

Throngs crowded onto ships and sailed around South America or tempted the sting of death by malaria or yellow fever in crossing the Central American isthmus. Some ethereal elation awakened an innate optimism that spread around the world. The gravitational pull of California was too much for people of many walks of life to try and counteract with sensible talk and quiet contemplation. It was a now-or-never moment, so off they went, following the siren song of the waters rushing off the slopes of the Sierra Nevada. Yes, many did strike it rich, but it usually came at the price of a tremendous amount of backbreaking work, long separation from loved ones, and the threat of life-shortening accidents and illnesses. Additionally, there were countless numbers who could barely hold still long enough to be counted by the 1850 U.S. census. They moved so quickly that there were very few of the same people in the same counties only a few years later.

One wonders whether a state founded on the principles of individual enterprise and profit could survive the eventual depletion of the "easy" placer mining locales, as corporate mining became the norm. This transition, from the very short-lived raucous gold rush to a more orderly development paradigm, is not often evaluated as the fluid process that it was within the various spatial scales in which it occurred.

The powerfully deceptive lure of riches for all who could find them is the simple explanation as to why so many people left their homelands and traveled thousands of miles to an unfamiliar land. Yet, there is something deeper, more elusive woven into the fabric of this and all stories of real lives. It is the longing for opportunities and places that are better in some way from one's present home and means of livelihood. Were the immigrants to California more opportunistic, more dissatisfied, or more desperate than those who stayed home? Without the ability to interview a large swath of these miners, it is impossible to tell how their thought processes differed from those they left behind. However, some ideas are revealed through reviewing the actions of the first California pioneers. This book follows the early Californians as they created their new society, and it examines whether they were successful in constructing a lasting landscape of prosperity and happiness.

PEOPLE'S PRODUCTION OF LANDSCAPE THROUGH TIME

There are four dimensions that intersect migration themes in the unfolding history of any place. First is the geography (landscape). The land is the base that people live on and use to make their livelihood. Although human and natural forces modify it, it retains a tremendous degree of continuity over time.

Second is the dimension of history or time. Discrete events can be contextualized within the larger framework of connected multiple events during the general period, which helps to explain their occurrence, ties them into the larger cycles of history, and underscores forward and backward links that reach far beyond the issue in question.

Third is the dimension of people, specifically, families. People are the actors upon the landscape, and individuals are tied together in time and space by their family relationships. Migration and marriage patterns, for example, are influenced predominantly by the historical period and familial factors (Macpherson 1984). Family genealogies are thus firmly entwined within both the temporal or historical dimension and the geographic or spatial dimension. In essence, the unveiling of a family history that spans numerous generations reaches out across space and time to reveal the greater historical trends of the world, as well as incorporating the friction of distance (or the inertia that often accompanies one's life-migration history) and the tendency of family members to cluster more closely over time compared with the general population (Bieder 1973).

Fourth and finally, is scale. Migration at the local level affects neighborhoods and other small locales. As scales shift upward to the community, scores of people moving in and out shape the destiny of the place. Ultimately, large regional structures evolve over time according to a complicated aggregate migration network of paths (transport routes) that move large numbers of people back and forth across a landscape of uneven, yet changing opportunities. Writers have produced a tremendous number of excellent works regarding this era and place in American history, but the rapid development of northern California has not often been approached from the multiscalar perspective that is introduced and explained in this book.

A MODEL OF MIGRATION AND SETTLEMENT

Combining the four dimensions of geography (landscape), time, genealogy (families), and scale produces a wide range of potential perspectives from which to make inquiry into the rapid settling of northern California. Each chapter of this book, although different from the other in terms of its migration dimension, is a piece of a related whole. To show these interconnections, I introduce a migration and settlement model derived from the specific historical geography of this region that could be applied to other places around the country. The model recognizes that the migration and settling process has had many similarities across different regions, but the rates and spatial manifestations have varied according to distinct geographic, economic, and temporal characteristics (fig. 1.2). It is loosely based on the concepts and models of others including Brown (1981), Meyer (1980), Muller (1977), North (1955), and R. Walker (2001a), as well as Bowden et al. (1971) and Turner (1953). California's developmental growth was rapid. Other regions of the country doubtless proceeded through the same general settlement process at quite different speeds. It is hoped that this model and approach may be useful to others studying historical settlement geographies in the United States.

The model combines the four dimensions to organize the analysis of

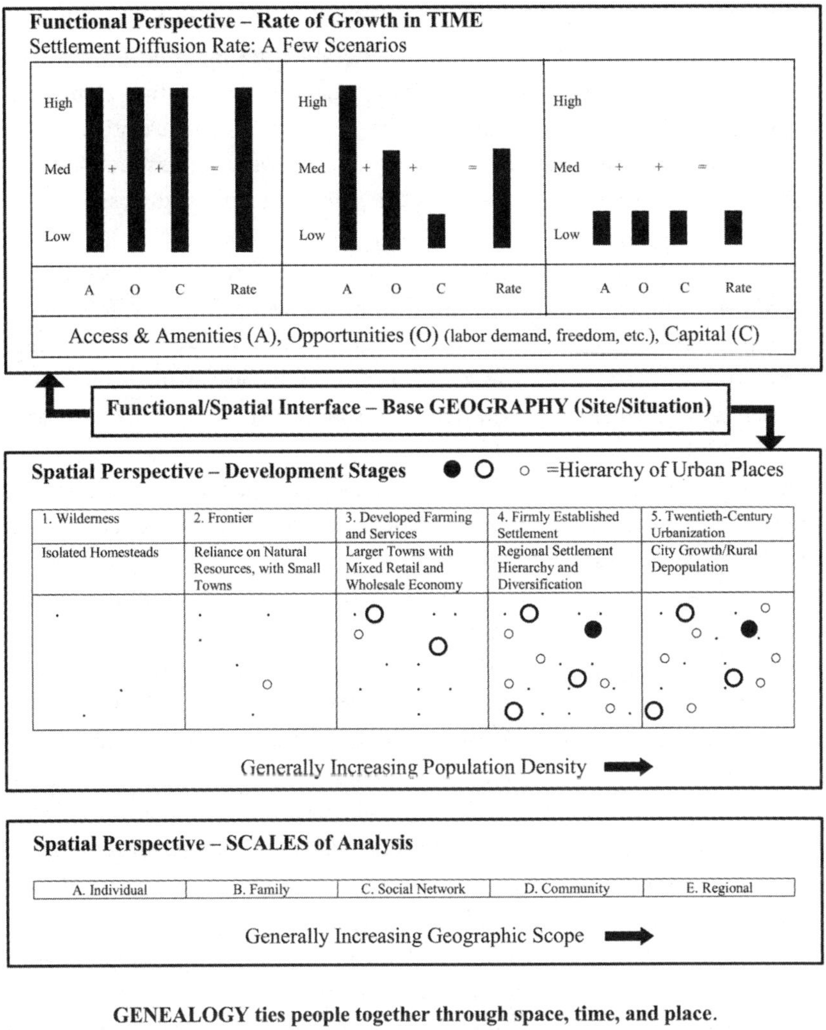

FIGURE 1.2. Migration and settlement model. Samuel Otterstrom, with design work by Jenny Jones.

migration and settlement change in various places. The *functional* perspective equates with the time dimension, meaning that migration is either rapid or slow through time depending on various factors. The *spatial* perspective focuses on settlement stages and the scales of analysis, which depend on the size of the migration and settling processes being studied. The *geography* (landscape) dimension is the interface between the two perspectives. Lastly, the *genealogy* dimension ties people together through space, time, and place, by blood and association.

Functionally in time, migration rates depend on the push and pull factors extant in both source origins and the migration destination. These factors include economic opportunity, amenities, access (transportation elements), social structure, freedom, and capital in the hands of migrants. Positive migration rates lead to settlement development, and regional expansion. Negative rates point to settlement stagnation and, in some cases, ghost towns, of which California has a large share.

The interface between the functional and spatial perspective is the geographic site and situation of the region, or base geography. The imprint of humans on the landscape, be it fast or slow, always occurs within the context of the extant geography. The outward reach, or diffusion of people and settlements is at first more reliant on natural resource availability, such as arable land, harvestable forests, productive fisheries, and mineral resources. Over time, the human occupance of the land changes, as manufacturing, services, and transport sectors grow on the original landscape and as greater population densities in core cities and towns occur.

Migration pathways facilitate or impede the speed and timing of the settlement of an area. These access routes are part of the geographic interface, and they greatly affect the rates of growth illustrated by the functional perspective. Initially, exploration followed by migration opens an area to future settlement. All places that have inhabitants must have some kind of access. The trails, roads, sea routes, and railroads that connected California to the outside world were the conduit for trade and for enhanced population growth. These pathways are still essential for the livelihood of the region, and their geographic placement and utility can channel growth to certain locales while handicapping it in others.

In the spatial perspective portion of the model, natural population growth and migration spur the development of settlements over a number of different stages that range from the individual farmstead to the complicated regional city-system. This progression through settlement development stages can be considered within five spatial scales. The first is the personal or individual scale that considers how seemingly unrelated people move and interact. Second is the family scale, which closely follows the individual scale. Migration and settlement patterns are greatly affected by extant and developing family relations. Third is the social network scale that creates cohesive groups of people due to religion, language, or other cultural characteristics. Whereas social networks may be dispersed or concentrated within the geographies of other culture groups, the fourth migration scale of community is placed within the specific confines of a bounded populated place. It results as individuals, families, and social organizations congregate and build economically diverse towns and cities. Finally, at the fifth migration scale is the development of regional economic and political relationships among many towns and villages in a specific area.

How do all of these pieces fit together? The genealogy perspective makes this more understandable. Its premise is that people at all scales, related both by family ties and social connection, are highly connected to their genealogical relationships in the way that they make their living and whether they migrate or stay in an area over time. Thus, the narrative explores the relationships of the individual to the community and to the region, always illustrating how the migration of related people across space and their economic interactions among various places serve as binding agents in regional place making. The movement of people across the land over time has resulted in marks they have put upon it through building, farming, and transporting. These marks, made in an endless attempt to live and improve living, are considered here within the various spatial scales of the varied geography of northern California during the last half of the eighteenth century.

SCALE, GEOGRAPHY, AND TIME IN CALIFORNIA'S PAST

I start at the most basic levels of geography, which are the individual to family scales, where migration tends to move toward preconceived "greener pastures" or a "promised land." It is a manifestation of the human desire for greater opportunity and locational improvement by migrating to a new place—be it a hoped-for Eden or Utopia—that will decrease personal poverty and heighten happiness. In other words, people use migration—often coupled with career shifts—in seeking to quench the burning desire for a better life. Alternatively, satisfaction with one's position in life, both socioeconomically and geographically, leads to a decrease in personal and family migration. Examining the lives of early Californians reveals the extremes of restless gold miner movement in the 1850s and 1860s followed by a more orderly migration flow in subsequent decades. The sheer volatility of gold mining communities in demographic structure and population persistence is a sure indicator that the argonauts had not found what they were looking for. Eventually, many of these miners settled down and turned to farming and other more traditional occupations as well as resuming a normal family life, often in other parts of California. Although the mines were the initial attraction and motivator, life as a "former miner" appeared to be more satisfying than continued toil and sweat in the mining districts.

Although frontiers were first mostly filled with men seeking their fortune, it was not long before women and children followed in corresponding numbers (Boeckel and Otterstrom 2009). These family pioneer migrants sought out a variety of opportunities in the West: farmers came because of low-priced or fertile farmland (hopefully both); would-be miners came because of fervent dreams of wealth; Mormons came to and left California because of religious ties and obligation. Otherwise, there was a great propensity for people to stay close to family homelands, especially during this late 1800s period. This was certainly the case in California. As more families moved to the Golden State, including fathers and husbands who had left brides and children to seek fortune and then returned to the East or other places to retrieve them, migration patterns also changed. Flows were tempered and constrained by family relationships, which focused and limited movement except when unusual opportunities or forces disrupted proximity as a general pattern. Over the following decades, family ties were imprinted on the landscape, with descendants of early Californians tending to cluster around the center of family settlement. Additionally, the migration between towns within California and the gathering from around the world into various communities is illustrated by both small- and large-scale genealogical analysis.

Moving up the scale, there are social networks that bind individuals and families together. Regions are often settled by more than just unrelated individuals and separate families. Due to their social cohesion, when religious groups move en masse to an area, they imprint distinct marks on the landscape because of their social cohesion. In this book, I focus on a religious movement that figures prominently in the history of California and of the greater West: the Church of Jesus Christ of Latter-day Saints, or the Mormons. From settlers of Yerba Buena (San Francisco) to laborers at Sutter's Mill at the gold strike, Mormons were on the scene very early on in the state's history. Their collective effect on the state is significant, and the intertwined relationship of this religious community with greater California history illustrates how historical geographies are often fashioned by groups of actors working in a coordinated manner.

Communities often develop through reliance on distinct mixtures of natural resources. When people migrate to a new area, they usually go with the purpose of obtaining gainful employment. The places that they eventually call home have varying numbers of jobs in primary, secondary, and tertiary sectors (in order: resource extraction, manufacturing, and services). Some end up nearly exclusively in mining, lumbering, or farming communities, such as the Sierra during the gold rush, or the rural farmlands of the Central Valley. In all these cases, the peculiar geographies associated with these resource-based economies helped directly drive the economies of countless towns in California and collectively stoked the burgeoning growth of the region's primate city, San Francisco (Brechin 1999).

The counties and settlements in the Sierra Nevada gold fields were extreme cases of single-resource dependence and fleeting prosperity. In concert with the exceedingly difficult task of tracking individuals in and out of the Sierra during the 1850s to 1880s, the towns and gold camps themselves came and went with an incessant boom and bust that defies a regular procession of settlement development. The Sierra settlement chapter (8) explores how different cities survived or failed in close proximity. Once again, geography mattered, for the places that prospered were on important transportation routes and cemented themselves as county seats or diversified their economies beyond the mining realm. Multitudes of other towns faded away when the gold played out and rapid outmigration decimated once thriving locales.

In contrast, Lake County is a less volatile example of how a community developed amid its particular internal and relational geographies. Lying to the north of the San Francisco Bay region and northeast of Santa Rosa, the center of the county contains Clear Lake, a large water body providing recreational opportunities for visitors. Fertile valleys with an amenable climate offered successful farming opportunities. And an active seismic foundation resulted in numerous hot springs, which laid the foundation for numerous health resorts. However, the mountainous ring that surrounded this pleasant place, although proving to be quicksilver rich, stunted the county's growth for many decades. Migration into this region was steady but not overwhelming. This combination of factors makes Lake County an excellent example of the many dimensions to economic growth in a diverse rural environment.

Although the concept is basic, it should be mentioned that just as individuals cannot be easily extracted from their families, and families cannot be separated from communities they themselves created, neighboring towns and cities develop in a highly integrated settlement system. Thus, to round out my analysis I explore the creation of ordered economic and settlement environment relationships among places. Across northern California, towns quickly settled into their own level of central place functionality through competition and geographic and historical fortuity. Outside of San Francisco, which commanded the economic fortunes of the region at the highest level, economic diversity appeared almost instantly in the fastest-growing towns, as growth begat more in-migration, more diverse manufacturers, and more varied service industries. It is astonishing how quickly San Francisco's hinterland became settled. These cities, built on natural resources, manufacturing demand, and service specialization, did not waste time developing and did not go through a long, multistage growth process. Instead, between 1850 and 1870, the relative size and economic diversity of the region's cities was already set.

The final chapter (11) ties all the scales together to illustrate how migration responds to regional socioeconomic realities. The reconfiguring of settled space through migration causes unending economic evolution in the pursuit of optimal conditions for happiness and prosperity, whose "best conditions" are always shifting according to time and cultural context. Tracking population change among all the northern California counties decade by decade reveals a picture of the ebb and flow of population movements from 1850 through the 1900s. The chapter shows how core, semi-core, and periphery counties often parallel each other in growth during different time periods. The analysis moves up to recent decades by showing the rural to urban migration that characterized the rise of manufacturing and later suburbanization. It also illustrates the boom and bust of resource-dependent mining, agricultural, or forestry counties.

Such movements of people are influenced by local, regional, and national economic conditions in pattern and intensity. Through these multiscalar relationships, some towns thrive, while others die, and the human impact on individuals and families is just as significant, although less noticeable on the ground than the changes in the built environment of places. Thus, I delve into regional, community, social network, family, and individual scales to show the interrelated geographies of migration and regional place making that have transformed California into the dynamic place and society that it is today.

2

Frictional Distance, Migration, and the California Transformation

Etched upon the landscape between the Sierra Nevada and the sea, are human markings that define the 160-year-old history of the state called California. Beneath such relatively recent impressions are those of the Native Americans, Spanish, Mexicans, and the ubiquitous natural forces that defined the stage before the gold rush era. Inaccessibility and remoteness from the rest of the United States may have slowed the population flow in the first years after the discovery at Sutter's Mill, but they certainly did not dampen the gold fever that spread around the world. Initially, thousands of men and a handful of women made their way to the mines, selling as much of their things as needed to buy passage on a ship or to outfit themselves to make their way across the continent. Did a dangerous tromp through the forests and wetlands of Panama stop them? No, although many died from fevers. Did many months of sailing around treacherous Cape Horn scare people away? No! Did seemingly endless days of traversing plains, deserts, and ascending high mountain ranges limit enrollment in one of the greatest gambles of the modern era? Hardly.

In the end, the migrational pull of gold, and to a lesser extent agriculture and trade, were more than strong enough to overcome such amazingly large frictions of distance and terrain. This chapter summarizes the geographic and historical factors that affected migration into California up until the transformative discovery of gold in a more condensed manner than more complete retellings by scholars such as Bagley (2010), Billington and Ridge (2001), Brands (2002), Roske (1968), and Starr (1973).

The scope of this chapter falls within the scope of the migration and settlement model (fig. 1.2). Functionally, the transportation barriers slowed immigration and settlement in California, but the economic opportunities that became apparent before the migration explosion of 1849 encouraged pioneers to overcome any challenge. Thus, settlement rates rapidly increased during this time. In earlier years, before gold ignited the imagination, in-migration was much more subdued in speed and volume, but the explorers and pioneers who

FIGURE 2.1. Early Spanish and southern overland exploration. Chris Maderia, Cassie Howe, ThinkSpatial, BYU Geography.

came first helped set the stage for the great influx of the mid 1800s. Migration and settlement scales during this pre-1849 wilderness stage were individual and exploratory or families that arrived after the area became accessible.

The Indians, Spanish, Russians, and Mexicans had made what is now northern California their home over the decades and centuries before the United States instituted its full-fledged push to the Pacific. These groups all made indelible marks on the landscape, many of which have persisted despite the rapid settlement and revolutionary reconfiguration of place and region that occurred in a few short decades after the onset of the gold rush. Interestingly, after its discovery, California was on the distant horizon of the Spanish, Russian, British, and American spheres. The Spanish approached from the southeast, the Russians from Alaska, the British mostly skirted it on their way to the coastal lands north of California, leaving the Americans to approach from the east and via long sea voyages. California, by virtue of its distance from Europe and the Atlantic seaboard of the United States, as well as its geographic boundaries of mountains, deserts, and the Pacific Ocean, was relatively untouched by these Europeans until the massive pull of gold fever shattered the settlement status quo.

It is inevitable that the first inhabitants of a place can more easily make lasting impressions on the landscape. However, even though Native Americans had the first and longest experience with the land, their impact has been mostly relegated to a lengthy list of place names and numerous small settlement sites that have given archaeologists much to puzzle over. Their lives are not well documented because of their lack of written records before colonization (Simmons 1998). Over centuries they had separated into many different bands within relatively small areas, speaking over a hundred different languages and numbering about 310,000 by 1849 (Simmons 1998, 48). The indigenous peoples subsisted on the bounty of the land, from salmon to acorns to deer and small game. But in their dispersed and independent tribal organization, they were unprepared to withstand encroachment by the advancing Europeans.

SPANISH COLONIZERS

First, and for the next three centuries, came Spanish priests and soldiers seeking to spread their Catholic faith to the Native Americans and to claim land for the mother country. To access this far-flung corner of the Spanish Empire, ships that sailed to the Pacific first brought explorers in the 1500s to the coast of California. In 1542, Juan Rodríguez Cabrillo pushed north from Mexico to explore as far as Point Reyes, followed by Sebastián Rodríguez Cermeño in

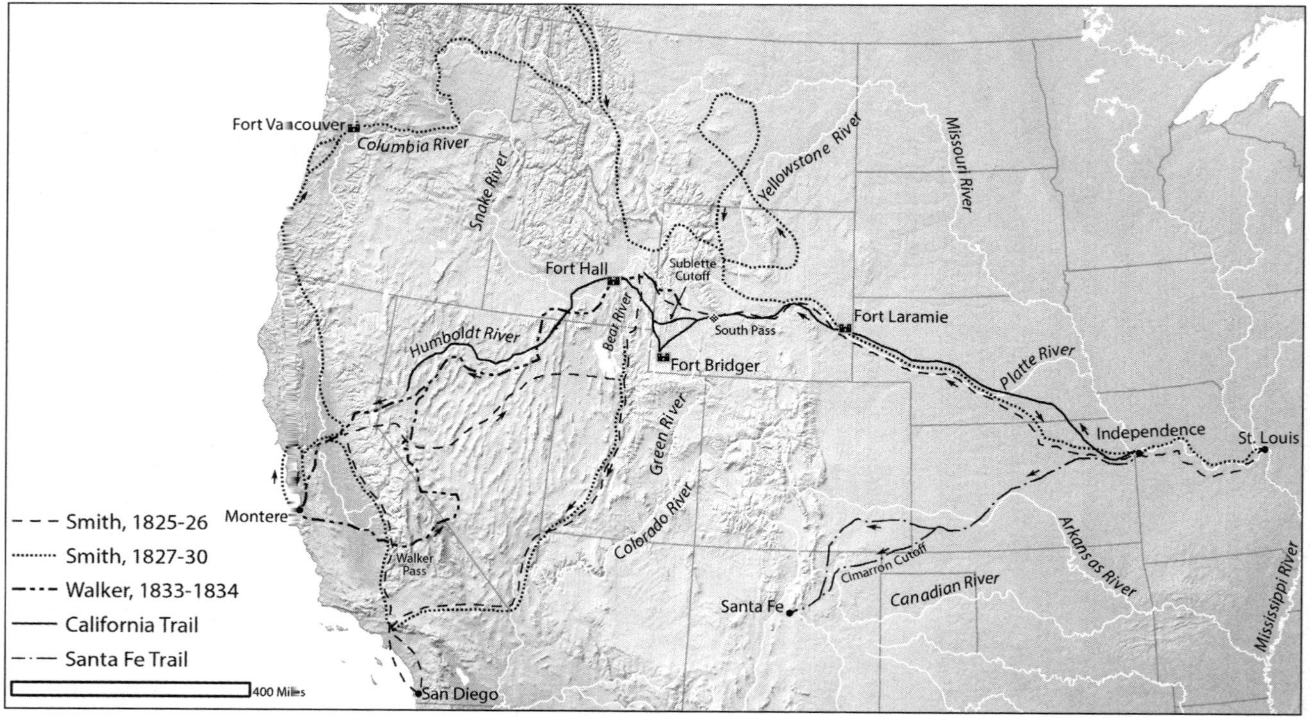

FIGURE 2.2. Western overland explorers and their routes. Chris Maderia and Jeffrey Stark, ThinkSpatial, BYU Geography.

1595 and Sebastián Vizcaíno in 1602,, who surveyed the coast in greater detail (fig. 2.1; Hayes 2007).

But it wasn't until the 1700s that the Spanish sent priests and others to actually settle California. Led by Don Gaspár de Portolá, governor of Baja California and Father Junípero Serra, settlers were sent from La Paz in Baja California to make the first permanent missions in Alta California. San Diego came in 1769, followed by the claiming of Monterey by Governor Portolá and Father Serra in 1770 (P. Johnson 1964). In Alta California, other missions were added along the El Camino Real, the trail that first connected San Diego with Monterey. This route became the lifeline of the twenty-one missions established in California by 1823, lining the Pacific coast from the south to north of the San Francisco Bay. The fact that California was very difficult to get to slowed its settlement: Alta California was literally months away from the nearest base in New Spain, and yet for years the settlers were precariously dependent on imported materials (P. Johnson 1964, 21–22).

Mexico declared its independence from Spain in 1810, and the demise of the California missions was set in motion during the confusion over their governance and secularization that began in 1833 (P. Johnson 1964). Mission lands were sold or given into private hands, but the population of Alta California was still fairly small compared with the onslaught of pioneers about to descend on this isolated region and who would soon take California away from Mexico. Yet the effective occupation of the California coast by these Spanish religious and colonial pioneers left a legacy of distinct Spanish colonial architecture that has been copied endlessly by modern architects. Additionally, the imprint of Spanish names on places, especially in the old mission realm, has been far reaching.

THE RUSSIANS: FORT ROSS

While the Spanish sought to hold central and southern California at the far end of its empire, Russia too dabbled in California. Making their way south from Alaska, the Russians discovered the potential income from fur trapping and food gathering in California. They established Fort Ross in 1812 north of Bodega Bay, but the venture did not prove profitable, and so they remained for less than thirty years. Once again, being at the very far end of the supply lines of a large empire, difficulty of access hindered settlement by outside peoples (Hayes 2007).

Neither the Spanish nor the Russians were successful in settling much more than the coastal areas of northern California. After independence, the Mexicans who followed the Spaniards designated and distributed large land grants for farms and ranches throughout many areas of California in an effort to settle this outpost of their young nation (Hayes 2007, 68).

OPENING BY LAND: MOUNTAIN MEN FROM THE EAST

What California required for growth was better transportation to improve access from the east and actions to promote settlement. The promoting began in small measure before 1848, but it was miniscule compared to what occurred after the find at Sutter's Mill. On the other hand, such land routes had already been developed over time by succeeding explorers and mountain men.

The two earliest explorers who crossed the Sierra in northern California from the east were Jedediah Smith and Joseph Walker. Their arduous journeys and Smith's death by Indians are immortalized in the Mountain Man literature. Jedediah Smith's party of trappers is also credited with rediscovering the South Pass through the Rocky Mountains in March 1824. In time, this pass became the funnel through which trappers and pioneers made their way to the west to Oregon, Utah, and California. During Jedediah Smith's adventures in the West, he and two partners actually blazed a route from northern California to the east in 1827. They had earlier entered California over the southern route in 1826 (Goetzmann 2006, 130–41). Unfortunately, Smith met an untimely end at the hand of Comanches along the Santa Fe Trail in 1831, and his route through the Sierra was mostly lost from memory (fig. 2.2; Carter 1982).

Joseph Walker, on the other hand, forged a path across what is now Nevada in 1833 that became one of the early routes for pioneers in the decade to come. He and his men approached the daunting Sierra in the cold and snow at the end of the year and barely made it over the mountains. Once on the other side, they found themselves in the breathtaking Yosemite Valley. They are credited for being the first Americans to make their entry into this spectacular area (A. Walker 1982). Walker and his company arrived at the Pacific some time later and then returned over the Sierra by a less arduous trail farther to the south. Subsequently their route became known as Walker Pass. This path was a notable artery into California but did not measure up to the importance of other trails to the north and south.

In the south, fur trappers explored a variety of routes to link Spanish New Mexico, based in Santa Fe, with the settlements of southern California. At first glance, these links to California seem less germane to our discussion, but the paths that pioneers traversed across this swath of land later brought their fair share of settlers, especially before the gold rush. Ewing Young and Kit Carson, among other trappers, set off from Santa Fe in 1829 and made their way west over the general route that Spaniards Juan de Oñate and Father Francisco Garcés had traversed in 1604 and 1776 respectively (Wilson 1965). Not long after Young's and Carson's adventure, William Wolfskill, an associate of Young, blazed the Old Spanish Trail from Santa Fe to Los Angeles during the fall and winter of 1830–31. This trail became the most famous route linking Santa Fe to southern California during the 1830s and 1840s (Wilson 1965).

Likely the most prominent non-Mexican to obtain Mexican land grants was the German-born Swiss John Augustus Sutter. He came to California in 1839 and was able to convince the Mexican governor Juan Bautista Alvarado to grant him a large swath of land in central California to establish a colony to cement Mexico's control in that region. The ranch was over forty thousand acres, and became a key center through the 1840s for Alta California and the focal point of the gold rush from 1848 on (Blodgett 1999). Sutter worked diligently to encourage migration to his settlement, and in 1845 sent forth the old mountain man Caleb Greenwood, who had arrived at Sutter's Fort in 1844 with the Stevens party, to encourage Oregon-bound migrants to come to California instead. Greenwood headed to Oregon and was successful, helping to facilitate and speed the settling of California before the discovery of gold. That discovery came at Coloma on the trail that first carried wagons over the Sierra in 1844, near Donner Lake on the Truckee route (Stewart 1983, 83).

Caleb Greenwood and his sons are important characters in this story, and functionally, although small in the annals of California's pioneer days, they serve as some of the spatial interconnectors among the dispersed geographic strands of this book. We find Caleb Greenwood opening the California Trail and drumming up "business" for Sutter specifically and California generally (chapters 5–6). Greenwood's son John was an early occupant of Greenwood, California (named for John or Caleb), which was smack dab in the thick of the early gold rush excitement. And while Caleb himself died in Coloma early in the gold rush days (chapters 5–6, 8), four of his sons were early settlers in Mendocino County, with the town Greenwood (now called Elk) being named after them (chapter 10). Caleb also was an early explorer of the territory around Clear Lake (chapter 9). Additionally, Caleb's son Governor Boggs Greenwood connects Caleb to Lilburn W. Boggs, former governor of Missouri where Caleb Greenwood spent some of his time between journeys up the Missouri watershed. Lilburn Boggs himself migrated to California from Missouri, just ahead of the ill-fated Donner Party in 1846 (whom Britton Greenwood helped rescue). He was also one that some Mormons in California probably tried to avoid because of some extremely unpleasant interactions in Missouri in the 1830s (chapters 4 and 5).

Caleb Greenwood's descendants spread across California and other areas of the West, an excellent example of the spatial process of multigenerational migration in the area (chapter 5). To wit, Caleb's great-great-great-granddaughter is the wife of this author, and was born in northern California herself!

EARLY WEST COAST CONNECTIONS

Before the gold rush, trade developed along the western coast of North America while these lands were contested by the British, Americans, Mexicans, and Russians. Meanwhile, the Native American population continued its existence but increasingly at the periphery. Part of this international contest amounted to trade competition along the coast. Under Spanish control, Boston "Yankees" exploited the sea otters until few were left. They then competed for cattle hides to take back east and sell for large profits. These Bostonians were more successful than the Russians or British, and after Spain lost control of California to newly independent Mexico in 1821, the Yankees increased their efforts and their pocketbooks. They would also bring manufactured goods from China, Europe, and New England to trade for hides and tallow (Holliday 1999).

The Willamette Valley in Oregon was in its early stage of settlement as well, and farmers there found themselves in need of cattle to improve their agricultural prospects. Philip Leget Edwards was teaching school at Campment du Sable in Oregon. He, along with Ewing Young, an experienced explorer, and William A. Slacum, a U.S. government agent, traveled by ship to Yerba Buena (San Francisco). They bought over 700 head of cattle from the disorganized Mexican government. The ensuing drive to Oregon lasted approximately four months from June until October. Some 630 of the 729 head of stock arrived there to boost the prospects of the Oregonians (Edwards 1989). This cattle drive served to further strengthen Oregon–California connections.

MORMONS BY SEA AND BY LAND

The first immigrant ship to arrive in Yerba Buena, after the United States gained control of Alta California in 1848 following the Mexican-American war, was the *Brooklyn*. The vessel brought some 234 passengers from the eastern part of the United States to San Francisco Bay. It took nearly six months, from February 4 to July 31 of 1846, to travel the harrowing route through both the Atlantic and Pacific Oceans. These migrants were distinctive in that they came to settle California in the name of freedom of religion, and not for purely economic reasons. As Mormons, they had planned to meet others of their faith who were to cross the continent by land following their spiritual leader Brigham Young. However, Brigham Young and his followers stopped at the shores of the Great Salt Lake in the Great Basin (L. Hansen 1988).

As the leader of the *Brooklyn* "Saints," the colorful and ambitious Samuel Brannan helped organize the migration and was instrumental in its successful execution, even though his leadership later became seen as self-serving. His business dealings in California were legendary, where he rose to be one of the wealthiest men in the state thanks to windfall profits related in large part to the gold rush. While he rose extraordinarily high, his fall back to earth was perhaps an even greater distance, as he died poor, single, and decrepit (Scott 1944, 441).

The voyage itself was an adventure. Mormons assembled from throughout the eastern states for the sea journey. After waiting for a snowstorm to disperse, they left New York for months of sailing. But just four days out, they encountered a powerful storm that pounded the *Brooklyn* so severely that Captain Abel Richardson told the passengers to prepare for the end. He said, "There is a time in every man's life when it is fitting that he should prepare to die. That time has come to us, and unless God interposes, we should all go to the bottom: I have done all in my power, but this is the worst gale I have ever known since I was a master of a ship" (L. Hansen 1988, 53).

Fortunately, after four days the storm abated and they were able to continue southeast until they were near Cape Verde. There they were able to take advantage of the northeasterly trade winds and head to the southwest, passing to the south of the northeastern extension of Brazil (Cape San Roque; L. Hansen 1988, 54). Upon reaching the equator they were met by the unwelcome calm of the doldrums. Luckily, they were becalmed for only a few days and were thereafter aided south by the southeasterly trade winds.

The warm breezes that propelled the *Brooklyn* through the wide expanse of the southern Atlantic eventually gave way to cooler climes. They were now approaching the very dangerous passage around Cape Horn. However, the ship had a relatively pleasant passage from the Atlantic to the Pacific side. Instead of fighting the relentless westerlies near the Cape, Captain Richardson sailed farther south (near 60 degrees south latitude) until "they encountered a south wind which carried them sufficiently west of the Cape where they then hauled to the north" (L. Hansen 1988, 57).

Another challenge awaited the *Brooklyn*. They had planned to stop in Valparaiso, Chile, to obtain supplies, but a second storm worked against them, blowing the ship the wrong direction. So instead, the captain decided to use the wind to head toward Juan Fernandez Island, of *Robinson Crusoe* fame. Although this gale was not as serious as the Atlantic one, it was responsible for the death of Laura Goodwin, pregnant, with seven other children onboard. She fell down a companionway in the unstable boat and died after experiencing complications from premature labor. The Goodwin family sadly buried their wife and mother on Juan Fernandez. On the island, much-needed water and food stores were replenished before the travelers headed northwest toward the Sandwich Islands (L. Hansen 1988, 58–59).

The ship glided efficiently to Oahu. Although a very busy place for sea trade of all types, "[t]he *Brooklyn* carried the first shipload of families intending to settle in California" (L. Hansen 1988, 62). There they unloaded freight, interacted with the natives, and met Commodore Robert F. Stockton, commander of the *Congress*, who informed Samuel Brannan and his companions that the United States was attempting to seize California from Mexico. This unsettled the Mormons somewhat, for they had embarked on such a monumental voyage with the original hope that they would settle outside of the United States to escape further persecution. Despite this concern, they decided to continue to California.

These migrants were now on the cusp of making an important impact on the early development of California. On the way from the Sandwich Islands to California, the *Brooklyn* passengers held military drills in case they were unwelcome upon arrival. When they finally sailed into Yerba Buena Cove, San Francisco Bay, on July 31, 1846, they were greeted by the U.S. Navy ship *Portsmouth* and learned that they were now in the United States (L. Hansen 1988, 62–65). The Mormons of the *Brooklyn* doubled the population of Yerba Buena, which changed its name to San Francisco soon after. Notable firsts for this group included the first school where English was taught, the first newspaper in San Francisco (the *California Star*, started by Samuel Brannan), the first agricultural colony established in the San Joaquin Valley, and a ferry that tied Fremont, San Francisco, and Sacramento together (Dunn 1997, 95).

A few short years later, the gold finds in the Sierra and their enthusiastic announcement throughout the world led to massive sea migration for a period of several years whose intensity is likely unmatched in history. The trip around Cape Horn was completed many times, even though the route was usually a challenge, as illustrated by the experience of the *Brooklyn*. The change of climate from the temperate latitudes in the Northern Hemisphere to the tropics and then back to the temperate latitudes of the Southern Hemisphere, with their inverted seasons, made it hard to stay comfortable. The travel through the calm doldrums in the tropics on one hand, and often-horrific gales in higher latitudes on the other, meant that the trip was anything but a holiday cruise (O. Lewis 1949). As soon as 1849, the advantages of a shorter route to California were apparent. This consisted of combined sea and land transport: sail to Central America, travel across Nicaragua or the Isthmus of Panama, and then find another ship on the Pacific side to San Francisco (O. Lewis 1949).

Trailblazing by the Battalion over the Carson Pass

At the same time that many Mormon Battalion members (e.g., several mentioned earlier who worked with James Marshall) and *Brooklyn* ship passengers were taking advantage of their proximity to the gold fields to enrich themselves through mining and trade (especially the illustrious Sam Brannan [Holliday 1999, 69]), some were preparing to head east in the summer of 1848 to meet up with Latter-day Saints in the Salt Lake Valley. They gathered in June in an area east of Hangtown (Placerville) that they named Pleasant Valley. Some forty-five men (and one woman) carved a route over the Sierra during July and August, known then as the Carson Pass Wagon Road (and now known as the Mormon Emigrant Trail), which was 9,550 feet at the pass. A substantial number of Mormon Battalion veterans traveled this way over the mountains and on to Salt Lake during the late summer and early fall (Owens 2004, ch. 5).

> Moreover the geography was favorable. Though they were moving upslope, they had a curious advantage of which they themselves were doubtless ignorant. The west slope of these mountains is furrowed by deep and difficult canyons, converging downslope like the branches of a tree uniting to form the trunk. Travel must be generally along the ridges. Only by luck, however, would anyone starting from the top happen to get upon the proper ridge.... But for anyone ascending the slope, from west to east, the ridges coalesce. The Mormons, therefore, having once got on a ridge, could follow it, by no virtue of their own, clear to the top. (Stewart 1983, 199–200)

This trail "became the most heavily used branch of the California Trail from 1849 to 1851, carrying tens of thousands of gold seekers directly into the heart of the northern mines" (Owens 2004, 160).

MASS MIGRATION

The miles passed so slowly and only the hope for a glorious discovery of easy riches pushed them on. The trails and ocean routes that led the early explorers, mountain men, and farmers to northern California were now being well worn by this flood of people. The gold rush drew unprecedented numbers of new migrants, revolutionizing landscapes from south to north. California's European population ballooned from about 5,780 in 1840 (Bancroft 1888, 3:699) to over 92,000 in the 1850 census.

On the other hand, the natives of this beautiful land were pushed aside. Their numbers decreased from an estimate of 100,000 in 1850 to fewer than 20,000 in 1916 (C. Miller and Hyslop 1983, 150). The land and people were changing at a rapid pace. There were so many individuals who came that we must sample these newcomers to obtain a sense of their complicated geographies. First we head to the Sierra (chapter 3) to find the early miners skittering about, and then we move west to the counties where many finally settled (chapter 4). These individual scale analyses will then give way to family and community scales to illustrate the interconnectedness of place and person over time.

3

From Movers and Shakers to Stayers and Builders in the Sierra Foothills

With the land paths imprinted in the dirt by numerous bold adventurers and the sea routes now followed more confidently by sailors from many distant shores, California was geographically primed for the settlement that exploded upon its north and the Sierra Nevada following Marshall's discovery at Sutter's Mill. The impact was immediate and profound. An internal stream of people flowed first to the Sierra from other parts of California in 1848, followed by an overwhelming cascade from distant population centers in the eastern United States and other parts of the world. The flood irrevocably changed the nature of California as a whole. The transition from a sleepy, predominantly rural outpost as a newly obtained U.S. territory to a state of considerable size and burgeoning population only took a few years.

This chapter explores the individual scale of the migration and settlement model and the rapid migration rates that typified the early period in the Sierra Nevada (fig. 1.2). The functional rates of growth were high because of the numerous mining opportunities and improved access that came with it. The influx of miners occurred during the first two development stages (wilderness and frontier) and started to wane when these mining areas transitioned to the third stage with more diversified urban development.

The many individual actors illustrate how transitory most of the players were in California during the gold rush. However, their singular impacts on the Sierra foothills combined to make a remarkable collective imprint through wholesale digging up and cutting down of the region's natural fixtures while establishing hundreds of gold camps, towns, and cities. Because of the fleeting nature of gold-mining success, many of these places faded away and nearly returned to their natural state following the gold rush. However, the miners substantiated Heisenberg's uncertainty principle: they went all over according to their personal decisions, but as a whole they acted both randomly (you can never quite determine where one particular miner would go) and predictably (e.g., percentages that stayed in the area, went back home, or left for another county). As we proceed to future chapters, this individual-scale geography will be incorporated within an ever-widening geographic construct.

MIGRATION'S SUDDEN IMPACT

California was not unlike other territories in being attractive to new settlers. However, the rate of influx was so high just before statehood that it exhibited a transitory demographic structure for numerous years until the rush of miners was finally followed by the more controlled flows of spouses and children.

A web of interrelated migration paths followed the initial call of men to the mines and gold-bearing watercourses of the Sierra Nevada. But it was not as simple as it might appear, for there were many streams of movement. Not only did newcomers flow to the newest mineral finds and then scurry on to greener pastures, sometimes they stayed and just changed occupations. For some, other more agriculturally promising parts of California beckoned with the promise of fertile farmland at reasonable rates and the return to their pre-California occupations. Many left the state and returned to families and homes in distant American locales. Still others left the country altogether for the mines of Australia, Canada, and elsewhere. One thing is certain: the early miners are exceptionally hard to pin to one place and track over time.

How did the miners lead to populating the landscape even though most didn't stay in one place long? How ephemeral were they, and how did gold form the basis for the great burst of growth that resulted in statehood in such short order? The point here is that mining functioned as a powerful yet temporary catalyst, while all the agriculture, industry, and trade that followed on the heels of the miners made California into a vibrant, sustainable economic society.

Place attachment can be very strong, but the early days of the gold rush show how land can, at certain times, have very little sticking power for impatient argonauts. In California, they found themselves wanderers on the earth, minds full of golden hopes that ever spurred their discontented beings toward seemingly more prosperous settings. Indeed, the key contributor to the fleeting nature of populations in the placer gold fields was the brief period of retrieval of an often meager quantity of the lustrous bounty. Thus, a connection to a place is weakened when local or even external means of support cannot be sustained.

It is not just to the characteristics of the miners that we must look, but also to their later lives and to their posterity. Their days were filled with activities meant to bring them happiness through economic survival and success. Their individual choices mingled with the environment, and the lives of their associates translated into either struggle or abundance. In subsequent chapters, we will examine how the children and grandchildren of these initial settlers spread across the state. As a whole, the various bits of many lives add up to a foundational layer of people who propelled California into an exceptionally vibrant state.

The furthest thing from the minds of new Californians as they hurried to mine, was the chance that their lives would actually be worse than in their prior homes. Perhaps the thought was there, but the overriding rush for success would not allow one to ponder the possibility of failure. Were the miners better off after they had labored so diligently to extract some gold flakes from the earth? What became of the miners once they arrived in California? How long did they continue mining?

These questions require multiple approaches to be fully addressed, but here we focus on patterns of persistence and volatility, origination and destination, and occupational starts and shifts within the mining regions. This was done by comparing census records from 1850 and 1860 of four Sierra mining counties and then matching records of those still found in California in the following censuses (1860 or 1870). The results illustrate the ephemeral nature of the mining population of the 1850s and 1860s and emphasize the difficulty of tracking miners from one census to the next.

The miners arrived, pushing across the imposing Sierra Nevada and down along valley and riverside routes on the western side. Beginning in 1849, these men came persistently, and the influx continued for many years. As explained, the Carson Valley route that wound its way down to Pleasant Valley and Placerville was made in 1848 by Mormons leaving California and heading east to Salt Lake City, but the following year the traffic was decidedly westward in nature. This stream of argonauts converged with others using different passes or arriving from Pacific ports. As they made their way into the gold fields, these amateur placer miners soon found that their new homes were the furthest thing imaginable from a stable place to live.

At first, most of the mining was of the placer variety, along the edge of the rivers and streams that flowed down the rugged hillsides. These sites were short-lived, as the hopeful miners constantly sought greener pastures.

It wasn't long until the gold locked up in quartzite rock was discovered. Although hard-rock mining had begun in Mariposa County in 1849, a strike at Grass Valley, Nevada County, by George Knight in October 1850 started this important innovation in the quest for gold. Shafts were sunk into the ground, and the metal was extracted more reliably when veins of gold-bearing quartzite were encountered. People could stay a little longer in these places than in the streamside camps (Nadeau 1965, 127–28). "Hardrock mines—requiring capital, machinery and steady payrolls—had given Grass Valley a permanent air in contrast to the boom-and-bust placer towns. It was one of the first cities on the Mother Lode to temper its rough male population with an influx of wives and families" (Nadeau 1965, 129).

TEMPORARY MINERS: HART AND CONE

The gold rush experiences of William Henry Hart and Gordon C. Cone are illustrative of the fleeting individual imprint so many California forty-niners made upon the landscape, even though their collective impact was monumental. These two men returned home after their time in California, which contrasts with many of the stories in the next chapter.

William Henry Hart (1829–1888) arrived from Illinois via the Carson Route, crossing the pass September 17, 1852. Along with his traveling companions, he came to Placerville on September 19.

> [W]e reached Placerville thus ending the trip. We found it a straggling irregular town of unpainted clapboard houses generally one story.... I believe I am the only one of the train who is satisfied with the trip. The rest all say that they never would undertake it again. The most of them seemed to expect a perfect pleasure trip, plenty of game, all fun, and no hardship. (Hart 1888)

A few days later, Hart was ready to try his hand at prospecting, and he left with eight associates on September 20 toward the South Fork of the American River (about 30 miles east of Placerville). They lasted a few days at their mining site before returning to town without any luck. Hart then made his way to Sacramento and San Francisco, returning to El Dorado County on October 9. He observed the operations at Coloma (or Colonia?) and tried his hand for a day there, making $2.60. Then he headed out to attempt shingle making on October 18.

> Commenced work as a shingle maker. Streeter and I attacked a huge sugar pine some four feet in diameter and felled it after a considerable hard work. We then took the cross cut saw and sawed the trunk into lengths of twenty inches, which we then split into blocks, and split these into shingles...October 23. During the last week the price of shingles has steadily fallen until they do not bring enough to pay our provision

> bills and we have concluded to try mining. Having sold our stock and tools we moved our camp utensils and provisions across the emigrant road and down a long hill into the bed of Weaver Creek (Weber Creek) where we had heard were good diggings.

They weren't that good. Hart continues,

> November 15. The recent rains having set the creek to running, filling all the holes and diggings and affording us water sufficient. Henry Miller and I set our cradles and commenced washing out the pile of dirt that we had thrown up. Finished the pile by the evening of the 17. The aggregate yield was $11.80 for 21½ days work making 55 cts a day to each man. Our provisions are costing us about 65 cts per day each.

They gave up on mining in that place on December 5, "went to town and sold the gold dust and determined to abandon mining in Weaver Creek." Next, Hart tried to find mining work at nearby Coon Hollow on December 11. In subsequent days, he also looked for work in Middletown, Cold Spring, and Gold Hill. The weather was quite bad through this time.

He finally left for Sacramento, arriving on Christmas Day 1852, and found it was suffering from a big flood. He then took a steamer down to San Francisco, staying at What Cheer House. After a few months, he decided to try his luck again in the mines.

On March 21, 1853, Hart headed to Mormon Creek in Tuolumne County. Here, he stayed and mined a claim until September 1, 1853. On that last day, he expressed, "[B]eing very much worn down with the labor and exposure of mining and in hopes of obtaining mercantile employment I have again concluded to try San Francisco." So in just one year, William Hart had lived in El Dorado County, San Francisco, and Tuolumne County, as well as other places in between. It is no wonder that very few miners in the 1850 census were even close to the same county ten years later when the next federal census was conducted.

Later, he briefly worked in San Francisco and then Petaluma, followed by Napa for a longer period. He left Napa on December 19, 1856, headed to San Francisco, and then took an adventurous sea voyage down to Nicaragua, across Central America and then to New York. The trip lasted thirty-four days, which was much faster than traveling across the continent by horse, wagon, or foot.

Gordon C. Cone, who lived 1807–1882, is another example of the instability of mining life (Cone 1850). He made the long and arduous journey from Waukesha, Wisconsin, to California, leaving Wisconsin on April 23, 1849, and arriving in Pleasant Valley on October 17. He worked with his friends for a number of months in the vicinity of Pleasant Valley from October to about Feb 13, 1850, when he summarized his frustration about the false advertising for California mining.

> Feb 12, 1850. It is a source of sincere regreat to me, that so much pains is taken by speculators to represent this country as affording such vast, and unlimited facilities for getting gold; for it is false. It is true, gamblers and speculators find great facilities for getting gold here in California, for they have the controll of the sustenance of the people, and compell them to pay such prices as their avarice may dictate, or meet death through the door of starvation. But the miners have no such facilities. Altho out of the immense multitude of people that are here for the purpose of mining, a few have succeeded well, yet the vast majority of them but little more than pay their way, so that in very many cases, it will take longer to save money enough to get home, than was calculated would be sufficient for them to get from ten, to twenty thousand dollars when they first started on their expedition.

He next spent a few weeks in Hangtown (Placerville) (from about February 13–March 1, 1850).

> Friday, March 1st/50. Our stay at Hangtown has been short. The diggins have been good here, but the best places have been dug out. There is a great many people here, and most of them are doeing little or nothing, as the wattr in the main creek is so aboundant, that the bars along its valley cannot be worked, and the side ravines have all been worked out. A very great many are leaveing for the mountains higher up, and all are intending to go as soon as the weather will enable them to camp out in a tent. We are now moveing our traps & things to Georgetown; a place farther north, and higher up in the mountains, and about 30 miles from here. It is a place where much gold has been found, but has been extensively dug over. In one ravine called the "Oregon Canyon" it is said that over one million of gold has been taken out since last fall.

His crew was about 4 miles from Georgetown for the remainder of his time (March 2–May 13), where they eventually had considerable success. However, in March, Gordon was struck with typhus shortly after selling his animals for $25 less than he paid for them and after writing in his journal on the night of March 9, "[T]he labor that I have performed, together with the care & anxiety that my mind is encumbered with, has nearly broke me down, as I have to go ahead in every-thing, & all responsibility falls upon me."

Cone was recovering by April 17, when he next wrote in his journal, but he was hit with scurvy in late April. His journal for April 22, 1850, outlines his thoughtful assessment of his success.

> Today the time of my men expires, Saturday being the last day of service for me. On that day, the two dug, & brought in two hundred & twenty Dollars worth of gold—this was doeing well for the last day. Dureing the past week they have had better success than we have had before in California. On Thursday of last week I discharged my nurse (one of my men that has not been in the mines for six weeks, on account of my sickness) which gave me nine days work, that brought me in almost five hundred Dollars. The three first days of the week I had but one hand in the mines, so that of what was taken dureing these days I had but half, as Williams (my partner) worked; but the three last days of the week he has not been able to work, so that all that was dug during this time was mine.
>
> On Monday they brought in one hundred & four dollars, & on Tuesday two hundred & twenty Eight Dollars. We have one piece that weighs seventy six Dollars, & one of sixty four (at sixteen Dollars pr. ounce) with many more that range from ten to twenty five Dollars.
>
> This however should offer no inducement for people to go to California, for it is among the "big strikes,"—and while they were doeing thus well, there was at least five hundred men at work within half a mile who, (with the exception of one small company) did not average more than two Dollars pr. day, many getting less, & some nothing at all. For two week previous to the week of good success, our Boys did not pay their board, & worked hard all the time. Thus it is, this California gold digging, you may dig for months and have but little encouragement, then again you may sink a hole where in a very short time you may take-out thousands. Thus it is, the gold when found in large quantyties is in small locations, & few in number, with no appearance on the surface to indicate its existance.

With finally enough strength to travel, on May 13, Cone began making his way back through Hangtown, Weberville, and Pleasant Valley to take care of his possessions and then on to Sacramento and finally San Francisco. It is obvious that he had had enough of California, and his description of the rapidly rising San Francisco on May 28, 1850, drives home that point.

> It is reported by many that the place is healthy, but my experience has taught me different, as well as hundreds & thousands that testefy to the

> same effect. Gambling, Drunkness & vice of every description is practiced here to an alarming extent. Thousands upon thousands are lost and win every day, and the practice of Gambling has become a passion with the people of San Francisco.

On June 1, Cone boarded the steamship *Oregon* heading for Panama. He arrived in New York City on July 8 and made it home to Wisconsin on July 16, 1850. Thus, Gordon C. Cone was in California for less than a year, returning to Waukesha in time for the census enumeration that occurred there on August 14. He died over three decades later in Clayton County, Iowa. He had lived in at least three different mining camps, a large part of which time he was ill. His story shows again just how incredibly mobile these early Californian prospectors were and that golden California failed to captivate all newcomers.

COUNTY STABILITY COMPARISONS

These two miners' stories give only a small snapshot of early miner migration. To get a better sense of the migratory volatility during this period, I chose four mining counties in the Sierra to study in more depth: Nevada, Placer, El Dorado, and Mariposa. An attempt to match people between 1850 and 1860 and 1860 and 1870 in the U.S. censuses was made, and the results follow.

Coloma, El Dorado County, was where Marshall discovered gold at Sutter's Mill. From there the cry of gold brought migration to the regions round about. The surrounding river valleys were soon swarming with would-be tycoons, making their temporary homes along the South, North, and Middle Forks of the American River and the Cosumnes River, as well as the tributaries in Placer and El Dorado counties. Communities that sprang up along Deer Creek and the southern tributaries of the Yuba River were important in Nevada County, while the Merced River system was a central destination in Mariposa County.

Transience was the rule. Mann (1982, 227) studied the communities of Grass Valley and Nevada City in Nevada County. He found that very few people stayed between 1850 and 1856. Only 8 percent of the farmers, artisans, and professionals or businessmen in Grass Valley in 1850 were still there in 1856, while about 24 percent of this same class of workers stayed around in Nevada City during those six years. The similarity between both communities was in the nearly complete turnover in the mining population in those few short years. Just 3 percent of the miners in each town stayed around for six years. Furthermore, Mann (1982, 266) reported that in Grass Valley and Nevada City only 4 and 6 percent, respectively, of the 1850 population was still in the same area in

1860. Between 1860 and 1870, the number of those who stayed only increased to 13 and 12 percent respectively. Rodman Paul sums up the itinerant propensity of the miners.

> If the miners had any one universal sin, it was their unwillingness to remain in one spot for a reasonable period of time. They were forever abandoning one district and hurrying off to another at the first whisper that rich ground had been found. In extreme disgust the *Sacramento Transcript* once denounced the people of its state as "certainly the most migratory on the face of the whole earth." On another occasion it compared the shifting mass of miners to "a sea whose tide knows no law—a rumor or a reality will fill a cañon with thousands of human beings." (Paul 1947, 84–85)

Assuming that volatility also reigned in other mining regions, analysis was conducted on four counties focusing on people who could be identified between decades. Only miners were sampled because of their assumed inability to stay put. Using the search features of Ancestry.com, the effort was to look for distinctly named miners in the first census year to locate in the next decennial census, specifically between 1850 and 1860, and 1860 and 1870. Other research (such as Mann 1982 and Dilsaver 1982, 1983) could only specify whether the same people were in the same county at each decade. Using Ancestry's search function makes it possible to locate many who left a particular county between decades. Only a small minority was found in each decade pair, especially after focusing on those who stayed in California after 1860. Tabulating the characteristics of the located individuals revealed patterns of birth location, occupational change, and place of residency.

Comparing the 1850 and 1860 decades was very difficult. Only Mariposa County was attempted for this period, due to the rarity of finding one miner in both censuses. Just 10.4 percent of the 539 sampled miners (in 1850) were found in the 1860 census. One can only imagine the poor census taker of 1850 trying to keep the placer miners on the different stream banks in one place long enough to get some kind of accounting of names, ages, and birthplaces. Only six of those 539 miners (just over 1 percent) were still in Mariposa County in 1860, while more than 35 percent had exited the state altogether. That left some thirty men who had departed the county but remained in California. Almost all of these people were in northern California counties, but only eleven of the thirty were still in mining (tables 3.1–2). Thus, the residents of Mariposa County in 1860 were very rarely those from 1850, while the prospectors who went elsewhere were not very likely still working for gold, whether in California or not.

TABLE 3.1. 1850 Mariposa County miners found in the 1860 census with birthplace.

Total Found, 1860 census	56	
Total 1850 Miners	539	
% Found	10.4	
	# FOUND	PERCENTAGE
Britain	9	16.1
Ireland	0	0.00
USA	46	82.1
Europe	1	1.8
Other	0	0.0
Total	56	100.0
	USA # FOUND	PERCENTAGE
New England	8	17.4
Mid-Atlantic	13	28.3
Southeast	20	43.5
North Central	5	10.9
Other	0	0.0
Total	46	100.0

TABLE 3.2. Location of 1850 Mariposa County miners found in the 1860 census and their occupations.

LOCATION	# FOUND	PERCENTAGE
Same County	6	10.7
San Francisco	2	3.6
Sacramento	2	3.6
Other Sierra	13	23.2
Other N. Cal. County	11	19.6
Other S. Cal. County	2	3.6
Out of State	20	35.7
Total	56	100.0

OCCUPATION	COUNT	PERCENTAGE
Laborer	7	12.5
Farming	17	30.4
Manufacturing	7	12.5
Sales/Service	9	16.1
Transport	0	0.0
Mining	11	19.6
Other/None/Unknown	5	8.9
Total	56	100.0

TABLE 3.3. Gender ratios (M:F) of select northern California counties 1850–1870 , derived from U.S. Census data.

COUNTY	RATIO 1850	RATIO 1860	RATIO 1870
El Dorado	28.4:1	3.2:1	2.2:1
Mariposa	51.3:1	3.7:1	2.6:1
Nevada	*	4.3:1	2.2:1
Placer		3.7:1	2.5:1
Yuba	43.7:1		

* 1850 – Nevada City 44.5:1; Grass Valley 17.2:1 (Mann 1982, 265)

The Chinese were even more challenging to follow. Census takers in El Dorado County in 1850 list about twenty people who originated in China. Ten years later there were 4,821 Chinese-born people in the county. It is nearly impossible to track this important minority between the censuses because of the common lack of two names (a personal name and a family name) and the problems of translating Chinese names into English characters for the census records. This is disappointing, because of the significant number of Chinese who migrated across the ocean to the gold fields during this period. Names such as How, Kiet, Hang, Cam, and On fill the El Dorado census sheets (and in 1870, the census takers gave each Chinese person the surname "Ah"). Consequently, it is most difficult to determine whether they stayed around for the next census, even with the knowledge that in 1870 the number of Chinese had dropped to only 1,552 living in El Dorado County (1,449 in 1880). Thousands had moved away, but again it is almost impossible to determine where they went. One can surmise that many returned to China or moved to places such as San Francisco (whose Chinese-born population ballooned from 2,693 in 1860 to 11,724 in 1870). The geography of these Chinese migration tendencies is decidedly complicated and barely understood.

Some forty-six of the fifty-six found miners in the Mariposa sample were born in the United States, pointing to the American character of the early gold rush. This changed greatly in the next decade where less than half of the located miners were native-born. Nationality aside, the numbers of miners who stayed in Mariposa County and continued in the mining profession had increased greatly. A greater measure of stability had begun to settle in. Something that probably helped add to this drop in migratory tendencies was the increase in the number of women in Mariposa, as well as in the other counties in the area (table 3.3). The overwhelming number of men in 1850 in the study region decreased greatly over the next decades so that a more even balance between the sexes became apparent, compared with the nearly nonexistence of women in the area in 1850. Still, males outnumbered females by over three to one in 1860 and by over two to one in 1870.

By 1870, a larger portion of the sampled miners from the 1860 census was located within the same county, and/or the same state as they were in 1860. The percentage of the 1860 miners located in the 1870 census ranged from 16.6 percent in Nevada County to 21 percent in Placer County. Even though many were still living in the state after these ten years, around 80 percent did not appear anywhere nearby. In other words, the total numbers listed (tables 3.4–5) are only those decisively matched between 1860 and 1870—others of the total were also probably in the state, just not conclusively located in the manuscript censuses.

TABLE 3.4. El Dorado, Placer, Mariposa, and Nevada County 1860 residents (mostly miners) found in the 1870 census and their birthplaces.

	EL DORADO	PLACER	MARIPOSA	NEVADA
Total Found, 1870	109	120	35	107
Total 1860	550	572	184	646
%	19.8	21.0	19.0	16.6

	EL DORADO		PLACER		MARIPOSA		NEVADA	
	#	%	#	%	#	%	#	%
All Birthplaces								
Britain	16	14.7	32	26.7	0	22.9	12	11.2
Ireland	26	23.9	24	20.0	10	28.6	17	15.9
U.S.A.	56	51.4	51	42.5	15	42.9	72	67.3
Europe	7	6.4	10	8.3	2	5.7	3	2.8
Other	4	3.7	3	2.5	0	0.0	3	2.8
Total	109	100.0	120	100.0	35	100.0	107	100.0
U.S.A. Birthplaces								
New England	14	25.0	11	21.6	5	33.3	19	17.8
Mid-Atlantic	12	21.4	21	41.2	2	13.3	17	15.9
Southeast	17	30.4	11	21.6	5	33.3	25	23.4
North Central	13	23.2	7	13.7	3	20.0	11	10.3
Other	0		1	2.0	0	0.0	0	0.0
Total	56	100.0	51	100.00	15	100.0	72	67.3

TABLE 3.5. Location of El Dorado, Placer, Mariposa, and Nevada County 1860 residents (mostly miners) found in the 1870 census and their occupations.

	EL DORADO (%)	PLACER (%)	MARIPOSA (%)	NEVADA (%)
Same County	51.4	52.5	48.6	60.7
San Francisco	9.2	13.3	11.4	4.7
Sacramento	4.6	2.5	0.0	0.9
Other Sierra	12.8	17.5	11.4	9.3
Other N. Cal. County	14.7	13.3	8.6	18.7
Other S. Cal. County	1.8	0.0	5.7	1.9
Out of State	5.5	0.8	14.3	1.9
Unclear				1.9

	EL DORADO		PLACER		MARIPOSA		NEVADA	
	#	%	#	%	#	%	#	&
Laborer	15	13.8	15	12.5	8	22.9	15	14.0
Farming	17	15.6	15	12.5	1	2.9	11	10.3
Manufacturing	12	11.0	11	9.2	1	2.9	11	10.3
Sales/Service	10	9.2	19	15.8	4	11.4	8	7.5
Transport	2	1.8	4	3.3	1	2.9	4	3.7
Mining	45	41.3	48	40.0	15	42.9	52	48.6
Other/None/Unknown	8	7.3	8	6.7	5	14.3	3	2.8

What were the destinations of approximately 80 percent of the 1860 miners who remain unlocated in 1870? Some had most assuredly died; others, after having been discouraged or enriched, returned to their homes in the eastern United States or overseas. Holliday (1981, 455) states, "Whether with pockets full of rocks or with the burden of failure, whether intent on coming back with their families or swearing to settle down in Maryland, they sailed from San Francisco each year—31,000 men (presumably no women) in 1853, 22,800 in 1855, 27,000 in 1858, 14,500 in 1860." This is certainly supported by our two examples of William Henry Hart and Gordon C. Cone, who didn't spend much time in the gold fields (less than a year in both cases), but instead found other lines of work or returned home after their brief foray into the mining occupation. Still others had moved on to different frontier areas such as Oregon, Washington, or Nevada. Another portion probably remained in California, but their common given names and surnames or changes in their reported ages, name spelling, or birthplace in the two censuses make matching them nearly impossible. These unknown migration destinations require a much more demanding matching effort beyond the manuscript censuses to bring the locations to light.

One reason many of the miners chose not to remain in California in these early decades was a general belief that farming in this region would be unsuccessful due to the dry summer climate. Holliday (1981, 354) states, "Farming seemed too slow of a route to wealth in 1850 and for several years thereafter. Besides, it required settling down, staying in California. If they were ever to take up farming again, it would be back home, not in the Sacramento Valley where it never rained from April or May until October or November."

A surprising comparison of the miners in both 1860 and 1870 in California and those who were no longer located in the state in 1870 is that proportionately more American-born miners appeared to have left California before 1870. The British and Irish, on the other hand, were the ones who were more likely to stay around. For example, nearly 42 percent of the 1860 Placer County miners located in 1870 were American, while some 62 percent of the nonlocated miners were born in the United States. Perhaps the migrants from across the ocean had less to return to or the added cost of returning to their homeland was too much (table 3.4).

There are two significant points here. First is that a large number, between 40 to 50 percent of those still in California, and around 10 percent of the total number of 1860 miners in these counties, had left their 1860 mining county of residence to settle elsewhere in the state. There they formed an important segment of the early population. Second, even though only a minority of the 1860 miners was found in 1870 in California, this number was much higher than that found between 1850 and 1860 in Mariposa County and parts of Nevada County (Mann 1982), meaning that the volatility of the 1850s was beginning to be replaced by some semblance of economic and demographic normality by 1870.

Thus, the migration of miners from the mines was still substantial, but more of these men could be found in the California vicinity than were located from the 1850 Mariposa miners in 1860. Additionally, a greater portion of those found in Mariposa in 1870 were listed in the same county in 1860 (Mariposa) in comparison with the 1850 to 1860 cohort: 10.7 percent in the same county between 1850 and 1860 compared with 48.6 percent between 1860 and 1870 (tables 3.2, 3.5). The 1860 to 1870 stayers in El Dorado, Placer, and Nevada counties were even greater in number, relatively, being the majority of those found between the decades.

Some 42.9 percent of the 1870 found miners in Mariposa County were still mining. Similar percentages of long-term miners were also found in El Dorado and Placer counties. Nevada County had an even higher portion of continuing miners at 48.6 percent. Additionally, according to Dilsaver (1982), many of those listed as laborers in the census were also working in mining. The miners in Mariposa County in 1860 were thus more likely to continue as miners into the next decade than the 1850 Mariposa miners. The same was most likely also the case in El Dorado, Placer, and Nevada counties. Interestingly, most of those who continued as miners from 1860 to 1870 stayed in their respective counties, although this proportion varied widely (El Dorado 82 percent; Mariposa 67 percent; Nevada 58 percent; Placer 83 percent). Along those lines, of those who remained within their 1860 counties, a majority were still classified miners in 1870, except for in Placer County (El Dorado 66 percent; Mariposa 59 percent; Nevada 68 percent; Placer 44 percent). The statistics illustrate the fact that among the individuals persisting in the same county, mining generally remained the career of choice for at least a decade of time. These changes emphasize the important shift in the nature of gold mining from a highly individualistic, rapidly revolving enterprise in the early days of the gold rush, to one with a more stable structure, often with laborers being paid for their time, rather than by the amount of gold they personally mined, although some miners continued working on their own. Quartzite mining was also instrumental to this increase in stability (Dilsaver 1983).

If half in the sample remained in the same counties during the 1860s into 1870, where did the others end up? Surprisingly few found their way to the economic and political hubs of northern California, San Francisco, and Sacramento. Additionally, even fewer ventured to southern parts of the state. Instead,

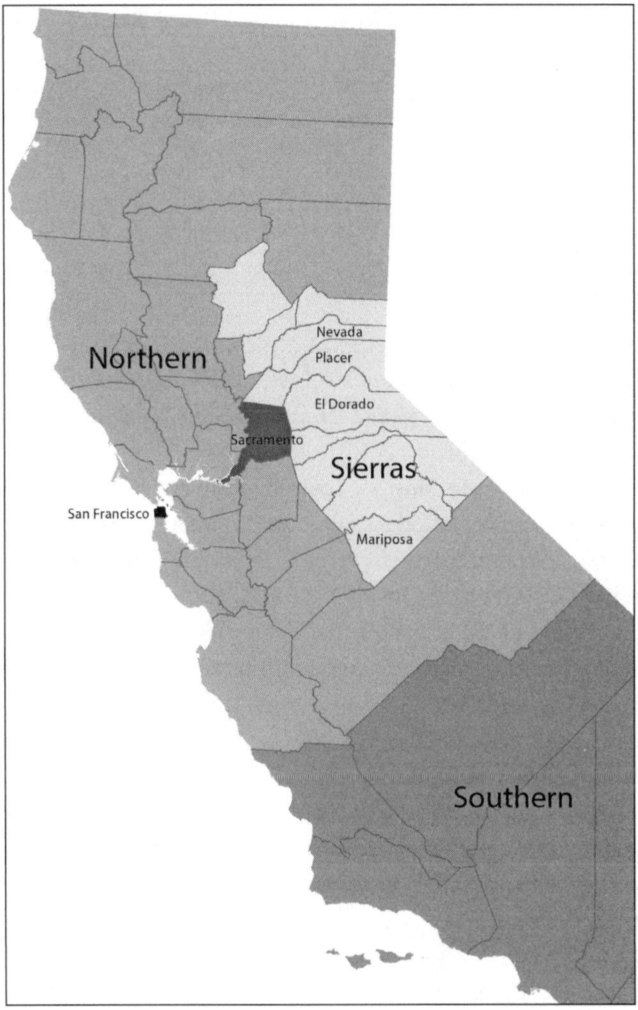

FIGURE 3.1. Sierra Nevada counties and other California regions, ca. 1860. Roman Huerta, ThinkSpatial, BYU Geography.

a more substantial number migrated to other northern California counties both in and out of the mining districts (table 3.5, fig. 3.1). This seems to indicate that the 1860 miners preferred less highly urbanized centers, whether or not they continued as miners. Furthermore, while many remained in mining, a wide variety of other occupations occurs in the 1870 census (table 3.6).

For example, John Driscoll, from Ireland, moved from Nevada County to Sacramento County between 1860 and 1870 and became a farmer. W. S. Cooper, born in Maine and a miner in Nevada County in 1860, was living in Tuolumne

TABLE 3.6. Occupations in Mariposa, El Dorado, and Placer counties, 1870 census.

	MARIPOSA (%)	EL DORADO (%)	PLACER (%)
Laborer	22.9	13.8	12.5
Farming	2.9	15.6	12.5
Manufacturing	2.9	11.0	9.2
Sales/Service	11.4	9.2	15.8
Transport	2.9	1.8	3.3
Mining	42.9	41.3	40.0
Other/None/Unknown	14.3	7.3	6.7

County in 1870 working as a civil engineer. Hiram Griffin, from Kentucky, was one who stayed on in Nevada County, employed as a quartz miner since at least 1860.

Placer, El Dorado, and Mariposa counties also had similar redistributions of people. James Langley left Placer County sometime in the 1850s and was a farm laborer in Sonoma County in 1860. David Bonner from Wales left mining and El Dorado County to try his hand as a furniture dealer in San Francisco by 1860. On the other hand, people like Charles Brogan from New Jersey, and Thomas O'Brien of Ireland, were found in Placer County both in 1850 and 1860 working as miners. All these California stayers, then, were the ones who would form a key constituency among the settlers of not only the Sierra counties but also much of the rest of northern California.

The gold rush drew a flood of people to California, but the mining districts didn't hold most of them for long. There were the James Langleys and John Driscolls, who turned to farming or other careers in different counties soon after their mining days, and then there were the William Henry Harts and Gordon C. Cones, who had enough of the Golden State and headed home far to the east. Yes, there were some who made a long-term economic activity from metal mining, but they were relatively few in number. As floods always go, areas filled with "fluid" miners until overflowing, and then they streamed away, leaving hundreds of struggling towns and deserted mining camps in the mining "floodplain." The volatility of the early years of the Sierra foothills gradually declined. The settlements that did survive either solidified or declined into ghost towns. In the next chapter, we will learn how many of the early miners helped build other areas of northern California, even as they continued changing locations. And in chapter 8, we will explore the factors leading to the rise, and often the fall, of the towns and villages of the Sierra.

4

County Pioneers and Their Many Migration Steps

From the previous analysis, it is clear that the 1850s and 1860s were times of great population volatility in the Sierra foothills. What was the lot of those who worked the mines and rivers for a short time and then went elsewhere in California? As a complement to the census searching of the previous chapter, another data source proved very helpful in answering this question.

County histories published in California during the late 1800s often included lengthy biography sections highlighting various community members. These "prominent citizens," who paid to have their personal particulars in the histories, left a very important data source (Fraser 1879). The biographies often include birth locations, migration paths, and occupations during the life course of the featured people. The men were mobile, much more so than one might suspect in an era of horse and wagon, sailing vessel, and later, the railroad. Total miles traversed in a lifetime spoke to the attraction of California as a destination and the remarkable efforts that went into getting there and finding a place to settle down after the gold fever wore off.

In terms of the migration model, functionally, migration rates continued to be rapid, even after the initial boom period in the mines (fig. 1.2). The rich geographic base that was much more than metal-bearing rock explains why the functional rate of migration was so high. Besides mining, new migrants found other economic endeavors, especially logging, agriculture, and basic services. Access from sea and land only increased over time, and the climate also facilitated the continuing migratory influx. The spatial scale for these migrations was at the individual level, but family connections are evident in the biographies. The lives of these migrants span the first four stages of settlement development from wilderness to a noticeable urban hierarchy within counties.

While migrants came from around the world, the bulk who appear in the histories of Sonoma, Mendocino, and Lake counties around 1880, for example, were from the U.S. Midwest and Northeast, followed by Germany, Canada,

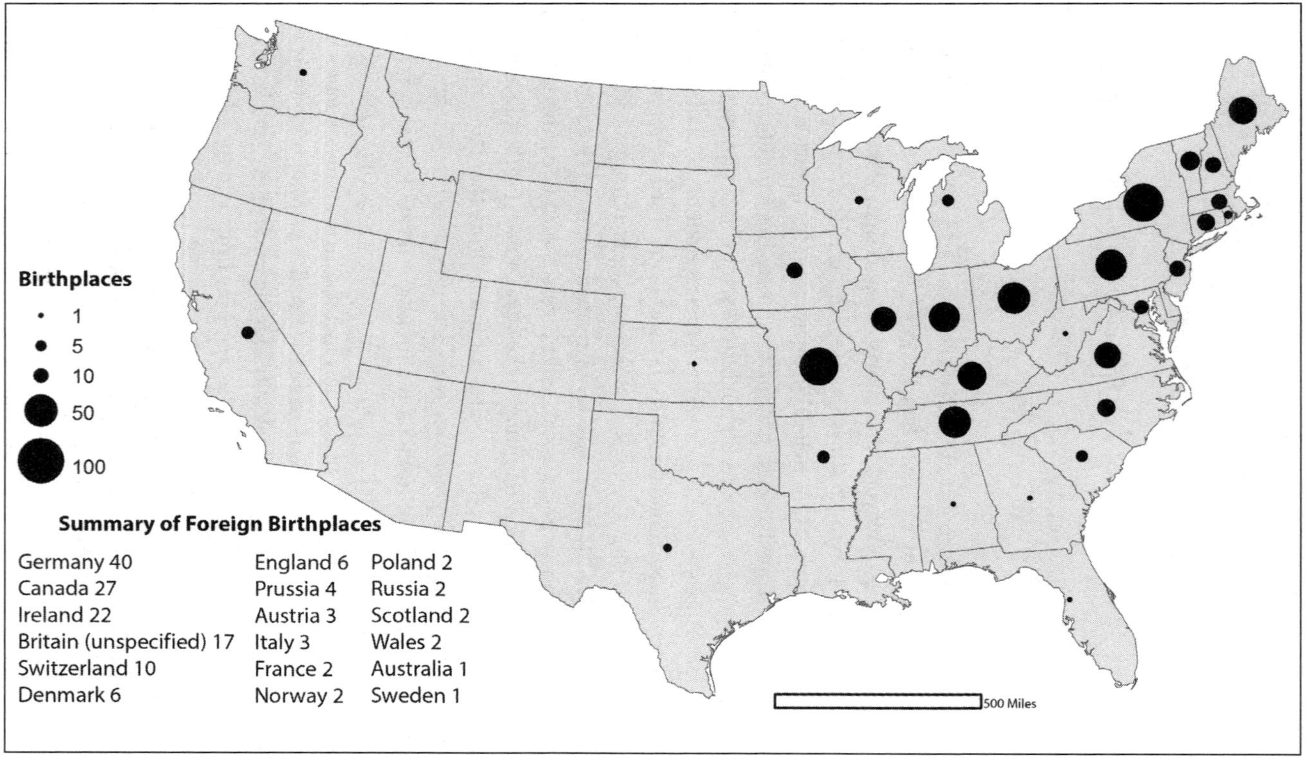

FIGURE 4.1. Birthplaces of migrants to Lake, Mendocino, and Sonoma counties (in county history biographies). Data sources: Munro-Fraser 1880b; Palmer 1880, 1995. Jeffrey Stark, ThinkSpatial, BYU Geography.

Ireland, and Great Britain (fig. 4.1). Interestingly, the southern United States is barely represented, which highlights a significant swath of the country that was not part of this great migration. The migratory tendencies of these men are worth exploration in and of themselves. Very few experienced a sedentary life, before or after they made their way to settle the "new" land of California. Censuses can only give a partial picture of these movements, because they were only conducted every tenth year. But in these cases, by having their comings and goings recorded by the biographer's pen, the men left a record that when reconstructed en masse illustrates a degree of mobility that rivals adults today in the United States.

Although they do not constitute a random sample of people in the community, the biographies do give detailed personal historical geographies of the lives of a certain set of men. These highlighted men represent only a small portion of county households. For example, just seven men from Denverton Township, Solano County, have biographies in the 1879 Solano County History, while there are some 56 Denverton households in the 1880 census. Additionally, Asians are rarely listed, so even with all the many Chinese laborers in Solano County in 1880, none are in the biography section. There are a few biographies of Hispanic residents—especially some who were in California several years before the gold rush.

I selected counties that were not in the Sierra gold fields to determine how the mining period directly involved men who later settled in other parts of the state. The analysis comes from Colusa, Marin, Napa, Mendocino, Sonoma, Lake, and Solano counties (Munro-Fraser 1880a, 1880b, 1994; Palmer 1880, 1881, 1995; Rogers 1891). These seven counties are located from the coast to the central valley, and as such display distinct economic diversity in the late 1800s (fig. 4.2).

INITIAL ATTRACTION TO CENTERS

One of Ravenstein's eleven laws of migration states that migrants going long distances focus on commercial and industrial centers (Grigg 1977; Ravenstein 1976). Or, in other words, "[T]he major direction of migration is from the agricultural areas to the centres of industry and commerce" (Grigg 1977, 43). In this research I focused on individuals who ended up in the cities, villages, and countryside outside the major center of San Francisco, so I cannot state how much stronger the flow was to the core versus the periphery. However, the massive growth of San Francisco during the mid to late 1800s is evidence of its importance.

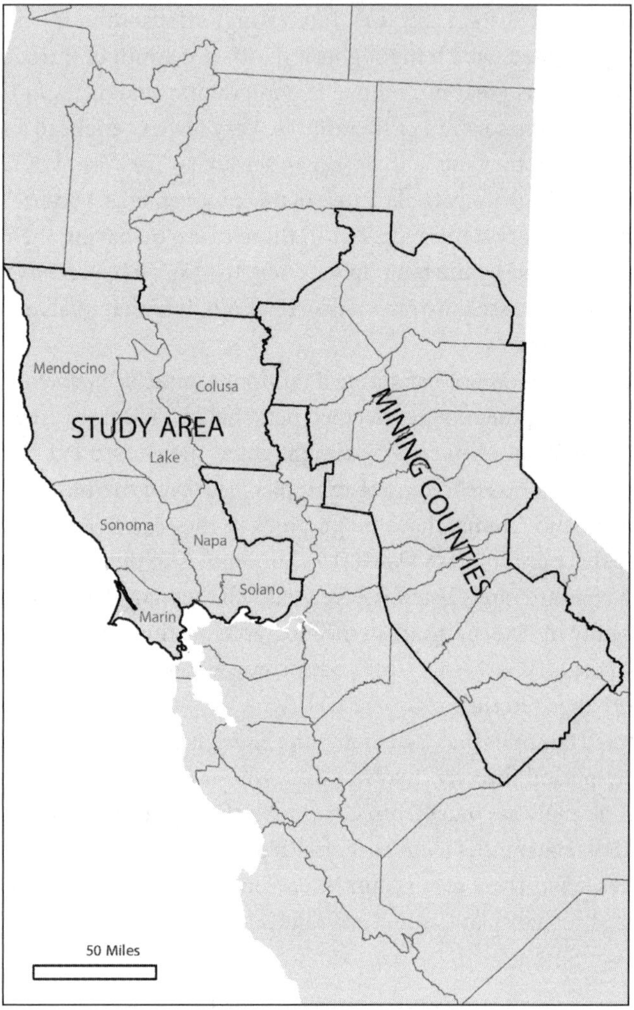

FIGURE 4.2. Northern California county history biographies study area. Ryan Shields, ThinkSpatial, BYU Geography.

The tendency to migrate to major centers of industry and commerce is certainly apparent for a good share of the early migrants to California. Because California was physically very separate from other U.S. population centers, migrants had to travel long distances to move there. Figures 4.3–5 illustrate the long distances that several of the men followed to get to California. What was particularly true for these early migrant cohorts is that soon after arrival they most often sought out the mining districts, the economic centers of the region. It then follows that in the northern California counties away from San

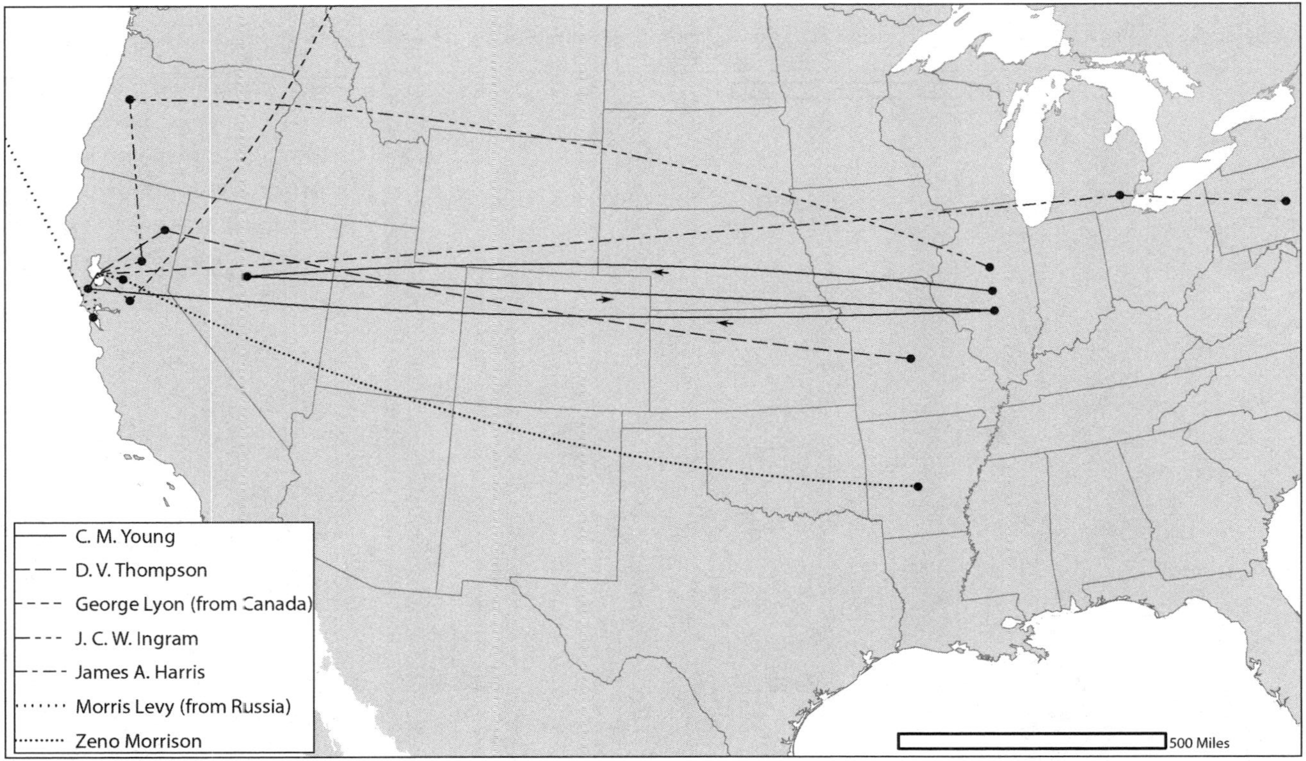

FIGURE 4.3. Migration paths of select Lake County migrants (through 1880). Data source: Palmer 1995. Jeffrey Stark, ThinkSpatial, BYU Geography.

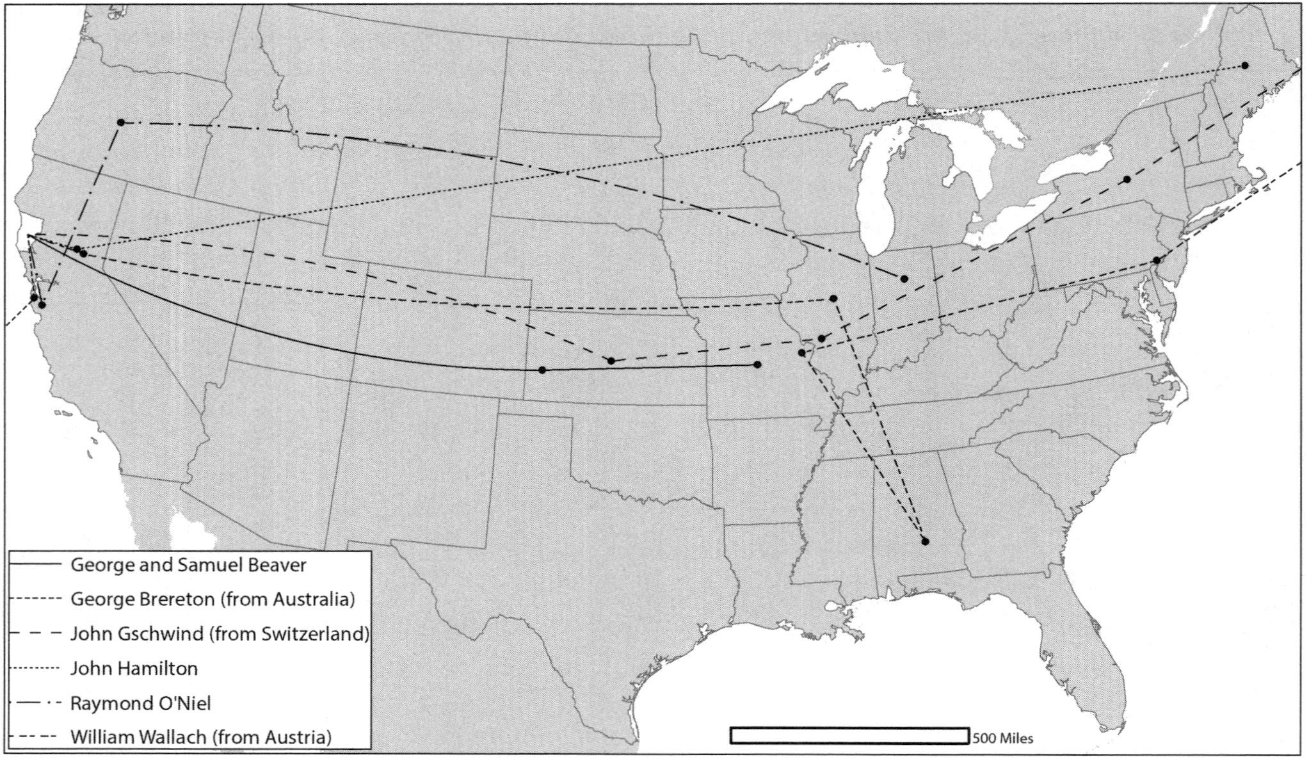

FIGURE 4.4. Migration paths of select Mendocino County migrants (through 1880). Data source: Palmer 1880. Jeffrey Stark, ThinkSpatial, BYU Geography.

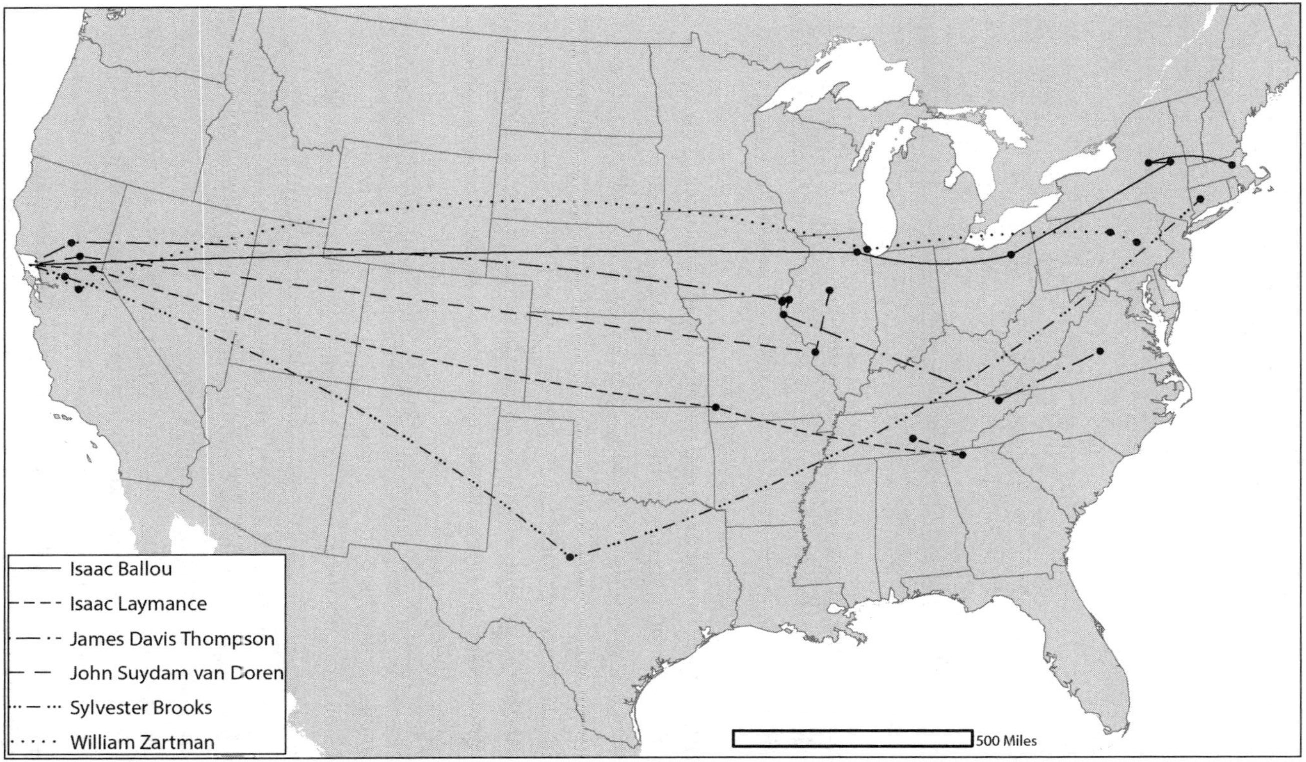

FIGURE 4.5. Migration paths of select Sonoma County migrants (through 1880). Data source: Munro-Fraser 1880b. Jeffrey Stark, ThinkSpatial, BYU Geography.

TABLE 4.1. First migration destination for residents of select northern California counties.*

	COLUSA	LAKE	MARIN	MENDOCINO	NAPA	SOLANO	SONOMA	TOTAL	% OF TOTAL
Mining	61	71	55	118	16	24	112	457	35.3
San Francisco	17	11	36	49	16	42	42	213	16.4
Colusa	32	0	0	0	0	0	0	32	2.5
Lake	0	15	0	0	0	0	0	15	1.2
Marin	0	0	64	8	0	0	3	75	5.8
Mendocino	0	0	1	37	0	0	3	41	3.2
Napa	1	11	0	7	25	0	3	47	3.6
Solano	0	5	1	4	2	32	7	51	3.9
Sonoma	1	11	16	26	2	0	70	126	9.7
Other	22	40	27	46	14	17	73	239	18.4
Total	134	164	200	295	75	115	313	1,296	100.0
% of Total	10.3	12.7	15.4	22.8	5.8	8.9	24.2	100.0	

*From biography sections of county histories, 1879–1891.

Francisco and the gold country, the largest number of men highlighted in the county histories of about 1880 came first to the Sierra foothills. Eventually, however, they made their way to the non-gold counties (table 4.1), if they did not go to a more permanent destination first.

Within all seven counties, 35 percent went to the mines upon arrival in the state. The counties farther from San Francisco (Lake, Mendocino, Colusa) had the highest proportion of miners at first entrance to California (at least 40 percent). Those closer to San Francisco were less likely to go to the mines initially (under 30 percent for Napa, Solano, and Marin). Of the men who arrived around the late 1840s and early 1850s, an even larger proportion made their way to the mines first. Therefore, although San Francisco was the largest commercial city, the mining center held more attraction in the early years.

Thus, the prominent citizens in the biographies generally did not move to San Francisco when they first came to the state. Most either headed to mining counties or other northern California counties (including the one they were living in when the biographies were written). Even in Marin County, which borders San Francisco on the north, only 18 percent of its highlighted inhabitants in 1880 had made their first California home in the big city, while 27.5 percent went to the mines first, and 32 percent came to Marin County straightway. In all the counties, the numbers that initially lived in San Francisco varied greatly from 6.7 percent in Lake County to 36.5 percent in Solano County. Even with this range, it is evident that San Francisco was usually less important as an intermediate destination than the gold districts or other parts of San Francisco's hinterlands for those who eventually settled away from the big city. Also, it shows that the geography of migration for those who found their way to the countryside often bypassed the core. San Francisco then, functioned as an economic primate city, but not as the prime population redistributor in the region of the study area.

Of course, many migrants spent brief amounts of time in San Francisco before moving to neighboring regions. Charles Quigg, an Irishman, left Ireland in 1845 at about fourteen years of age, probably to escape the famine there, and settled in America. He apprenticed as a boilermaker in New York City before sailing to San Francisco in 1851. Although San Francisco was his initial stop, he moved three to four weeks later to Benicia, Solano County, where he worked for the Pacific Mail Steamship Company. In Mr. Quigg's case he used San Francisco as a waypoint to a more permanent residence (Fraser 1879, 447).

The great number of these men, who first headed to the gold districts of the Sierra before moving on, had varying levels of success in the gold digging enterprise. What is perhaps different or distinct about the California experience is the great restlessness of the early migrants. The fact that the mining regions of California were large, and the different towns therein were separated by significant distances, meant that even if one worked as a miner longer than a year or two, he often traveled to various camps and mining towns over time hoping to increase his luck, as illustrated in the previous chapter. The fleeting and sometimes illusory success of mining at specific sites in the foothills translated into an unstable center that attracted thousands from around the world but could not retain them for any period of time.

In Lake County, W. W. Greene is a good example of the long distances and multiple moves that many California pioneers undertook. He was born in Missouri in 1827 and came to California in the wake of the gold rush in 1850. He mined for only a month in El Dorado County before going into merchandising, then farming in regions round about (Placer County, Sacramento area, Rough and Ready, Nevada County, and Auburn). In early 1853, he returned to Missouri via Panama in order to retrieve his family. After a long ride across the overland trail, he and his family arrived back in Auburn, California in October. He remained there as a hotelkeeper and teamster until he moved to Colusa in 1858. There, he was a stockman, hotelkeeper, and merchandiser in the city and county until 1869. He then moved to neighboring Lake County and was a hotelkeeper and merchandiser there. He also ran the Bartlett Springs resort for a time, as well as being the proprietor of Greene's Hotel from 1872 to at least 1880. Lake County was an important tourist destination by that time, as chapter 9 will underscore (Palmer 1995).

If a person happened to come to California later in the 1850s or into the 1860s, the mines would not have the same attraction, unless that person moved

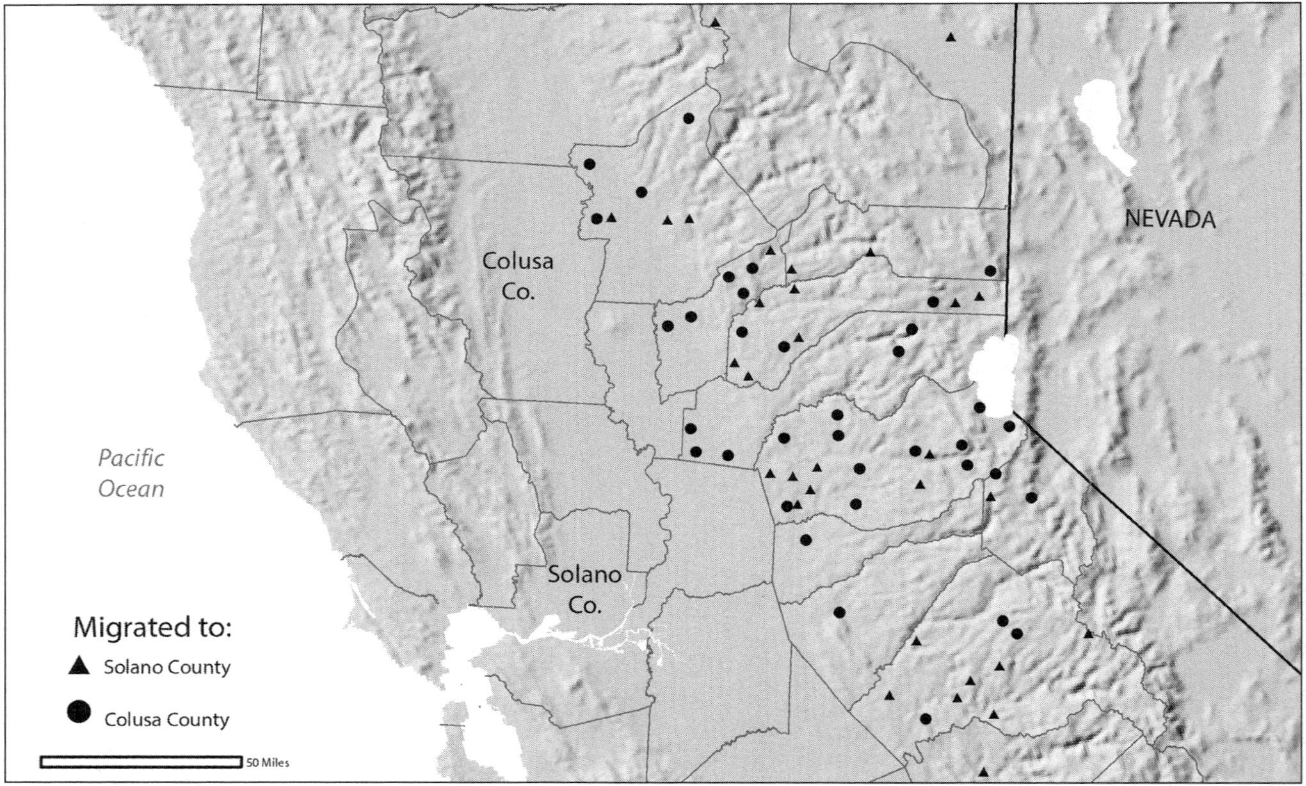

FIGURE 4.6. Mining locations of men who later moved to Colusa County (by 1891) and Solano County (by 1879). Data sources: Fraser 1879; Rogers 1891. Roman Huerta, ThinkSpatial, BYU Geography.

directly to the mining areas and stayed in the vicinity. Take D. W. Walker for example. He was born in 1823 in Pennsylvania. After doing some work in the lumber business, he moved to Wisconsin in 1856, and worked as a blacksmith and a merchandiser until he migrated to Santa Clara County, California in 1860. He farmed there until 1865 and then moved to Sonoma County and then farther north to Mendocino County. Nowhere in Mr. Walker's biography is there mention of time spent in the mines, which is quite different from records of many other pioneers who came to California about a decade earlier (Palmer 1880, 617).

In contrast, Henry Willard, also found in Sanel Township, Mendocino with D. W. Walker in 1880, was one of the early volunteers who came with Colonel Stevenson of New York to San Francisco in 1847. News of gold in the Sierra lured him to Placer County in 1848 for a year. He then made his way back from the mines to San Francisco, then north to Marin County. He migrated to Mendocino County in 1856, about eleven years earlier than D. W. Walker. Willard's mining experience was an important difference distinguishing the two men from each other.

AFTER THE MINES: SHORT-DISTANCE MIGRATION

If one wonders where all the people went after they left the gold mining districts of the Sierra foothills, a significant number of the former miners can be found later in the districts west of the mountains, such as Colusa and Solano counties (fig. 4.6). This is a frequent pattern in the lives of the highlighted men of these places.

Another of Ravenstein's laws supports this pattern closely: the majority of migrants move only a short distance (Grigg 1977). The amount of moving that went on around California, for instance, was significant but usually occurred in small segments. The efforts to procure a living meant that if gold wasn't coming out of the earth in one location, a man had to get somewhere else to find profitable work (table 4.2). Another move that usually occurred in these instances dealt with occupation shifts from mining to other common trades. These new work incarnations included teamster, farmer, rancher, merchandiser, and hotelkeeper among many others. It is surprising to see how many different careers one individual could try in his life, rivaling what many people experience in the twenty-first century. Migration and occupation change in early northern California history were inextricably connected.

Because many of the men made their way to California alone, it was common for them to return east briefly to gather family members (table 4.3).

TABLE 4.2. Lifetime moves of residents of Colusa, Marin, Napa, and Solano counties, based on county history biography sections, 1879–1891.

NUMBER OF MOVES	NUMBER OF PEOPLE	%
Colusa		
11	3	2.2
10	0	0
9	2	1.5
8	4	2.9
7	9	6.6
6	10	7.3
5	15	11.0
4	40	29.2
3	22	16.1
2	22	16.1
1	10	7.3
Marin		
13	6	2.9
12	2	1.0
11	2	1.0
10	2	1.0
9	6	2.9
8	11	5.4
7	20	9.8
6	18	8.8
5	26	12.8
4	31	15.2
3	36	17.7
2	35	17.2
1	9	4.4
Napa		
9	14	18.9
8	6	8.1
7	5	6.8
6	10	13.5
5	12	16.2
4	12	16.2
3	8	10.8
2	6	8.1
1	1	1.4
Solano		
16	1	0.9
15	1	0.9
14	0	0
13	1	0.9
12	3	2.6
11	3	2.6
10	6	5.2
9	5	4.3
8	3	2.6
7	5	4.3
6	13	11.2
5	23	19.8
4	14	12.1
3	17	14.7
2	16	13.8
1	5	4.3

TABLE 4.3. Number of men who traveled east after moving to California (and later returned to California).

COUNTY	# OF MEN	% OF TOTAL MEN IN BIOGRAPHIES
Lake	34	20.7
Mendocino	84	28.5
Sonoma	102	32.5
Colusa	11	8.0
Marin	18	8.8
Napa	16	21.6
Solano	22	19.0

Derived from county history biography sections, 1879–1891.

The magnitude and expense of such an undertaking, even after the Transcontinental Railroad, is impressive. Yet, some of the men even made the trip multiple times. It certainly took a lot of incremental movements for the men to settle down in one location.

Born in Hanover, Germany, August F. Mahlmann had seen the world by the time he arrived in San Francisco in 1852. A baker by trade, he had left his apprenticeship at age seventeen to travel. And travel he did. He began at the busy port of Hamburg on the Elbe and proceeded by sea and by land, whereby a wide range of cities and countries came to know his feet. Stralsund on the Baltic, Berlin, Danzig, Riga, Moscow, Hungary, Turkey, Tricase (Italy), Venice, Rome, Bern, Basel, Metz (France), and Paris, among other places, were all on his itinerary over the next nine years. Unfortunately, August became involved with Swiss nationalists and was wounded and imprisoned in Prussia for over a year. When he was kicked out of Prussia, he found his way to New York City in 1850. He made his living there with his skill in bird stuffing, obtained some years before in Hamburg. However, the lure of California finally came. The attraction of the mines induced him to leave by ship via Panama to the land of opportunity. One can only wonder how August compared the raw mining districts surrounding Nevada City to the great cities of Europe that he had frequented years earlier. San Francisco, although growing quickly, was no more than a child in comparison with those mature metropolises.

August Mahlmann followed the lead of so many others, digging for gold quickly and getting out of town with any profit he could muster in one year of work. His takings must not have been large, because he found new employment in the sawmills of coastal Mendocino County. At age forty-six, after another eight years in the forest products economy, he moved to Little River (on the Mendocino coast), built a hotel, and married a Danish woman (Palmer 1880, 548). Mahlmann listed his hotel enterprise in the *Paulson's Directory* of 1874. His vantage point had been radically transformed from globetrotting German to sedentary hotelier on the edge of the disappearing redwoods on the rugged northern California coast.

This is a seemingly standard pattern of initial attraction to the core mining center followed by migration to the core or semiperipheral area, in some cases continuing to an even more peripheral endpoint. Many men also originated in more rural areas. A fascinating example is Lilburn W. Boggs, who hailed from the relatively young state of Missouri. Boggs was the governor of the state of Missouri in the 1830s and was also a merchant when he was not politicking. He continued pursuing various political aspirations until 1846 when he decided to leave Missouri for opportunity in the West. He migrated to California as part of the Russell-Boggs party. The Donner-Reed train joined with the Russell-Boggs group for part of the journey. Fortunately for Lilburn Boggs, and those following his lead, he decided against taking the Hastings route that the majority of the Donner Party followed. Instead, he returned to the main trail with his group and arrived in Sacramento on October 28, 1846, without incident and without much historical recognition. Thus they beat the big snowstorm that stopped the Donners, Reeds, Murphys, and others high in the Sierra. Those poor families had taken the "shortcut" that was anything but short, and the tragedy of the Donner Party, in contrast to Lilburn Boggs's contingent, is known far and wide (Marriott 1979; Nelson 2011).

In the spring of 1847, Boggs purchased a store in the town of Sonoma north of San Francisco (in present-day Sonoma County). He was appointed alcalde, for all of the Northern District of California. (An alcalde was the highest civil authority with legislative, judicial, and executive powers [Marriott 1979, 78–79]). His business benefited from the discovery of gold in the Sierra, and he did well for a number of years. Around 1852, he left public service and moved to his farm in Napa, where he lived until his death in 1861 (Marriott 1979, 90). Thus, Lilburn's migration steps included the major move to California, and then shorter relocations to Sonoma and nearby Napa. He did not work in the mines, but his merchandizing and governmental appointments clearly benefited from mining. Missouri was a common origin for migrants, though one was much more likely to find an unknown farmer or miner among them than a former governor such as Lilburn Boggs.

The process of migration into California is distinct in that many of the individuals, even with short biographies in the county histories, were prone to a zigzag relocation pattern that occurred for ten years from the start of the gold rush until about 1860. They moved back and forth between the mines in the Sierra and the counties to the west, or they would take at least one journey back east to regroup and return to California with family.

Lesser-known characters than Lilburn Boggs abound, but their stories are no less illustrative. Take for instance Amos Snuffin. He was born in Ohio, and his family moved to Missouri about 1842. In April 1850, he came west to seek fortune in the mines, arriving in Hangtown (Placerville) that summer (Palmer 1880, 668). He then began the back-and-forth series of short migrations common for the period. He moved near the coast in Sonoma County in 1851 to work on a farm, and then he returned to the mining area in 1852 for six months. Snuffin then returned to Sonoma County until 1857, then moved north to Mendocino County, where he pursued various occupations such as farming and wool growing on 710 acres near Ukiah to the time of his short biography in 1880.

Another mover was George E. White. He was born in 1831 in what is now West Virginia. He came in 1849 to Shasta County to try his luck mining but soon moved to the Sacramento area. In 1850, he was working in a store in Nevada City, and later in the year he became involved with freighting based in Sacramento. It was just two years later that he returned to West Virginia via Panama. He next headed to Missouri in 1852 and then on to Round Valley, Mendocino County, in 1854 to prospect for gold, where he probably didn't find much. He went back to his home state of West Virginia in 1868 and married. In 1872, he moved his family to Round Valley where he remained until the biographer came through in 1880 to record his life. The county history presents a positive picture of George White, but the truth is that he became a wealthy land baron in Mendocino County who used merciless techniques to increase his holdings: "In the isolated and spectacular country of the Yolla Bollies, George Edward White was determined to build a huge cattle empire, and he ruthlessly set out to do so. From his headquarters in the southern part of Round Valley… he gradually acquired over thirty-five thousand acres of the best range in Mendocino, Trinity, and southern Humboldt counties" (Carranco and Beard 1981, 217). Space limits further retelling of his exploits.

Some came west with the rush but became engaged in something other than mining. In 1814, somewhere in the green hills of eastern Tennessee, a little boy was born to Harris Standley and Rachel Olderson. His dad thought it fitting to name his new male child after himself, Harrison. His parents were transplants from South Carolina and Virginia who—if the story is true—were heroic pioneers in their own right. In her strongest years, Rachel could "shoulder three bushels of wheat" while standing in a half-bushel measure (Palmer 1880, 617), and Harris supposedly lived to the ripe old age of 115. Perhaps these solid characteristics help explain the family's wanderlust, because Tennessee soon lost its appeal. Around 1817, they moved west to Howard County in central Missouri territory to farm. In 1834, Harrison Standley married Elizabeth Shelton, who was originally from Buncombe County, North Carolina, and they had their first children in Clinton County, Missouri, north of Kansas City. The growing family moved to nearby Andrew County in the late 1830s, and more children were born. Before 1850, Harrison busied himself as a teamster in St. Joseph, Missouri, for a time.

Harrison must have heard enough of the glowing reports from California to induce him to leave his established home and head west to the bustling gold center of Hangtown (Placerville), California, in 1850, right after the initial gold rush. However, instead of panning for treasure, he continued his trade of teaming for a few years. He then returned to Missouri via Panama to gather his family to cross the plains together. They bypassed the mines, taking up residence in Petaluma in Sonoma County. He spent his time hauling goods until 1858 when he moved to Ukiah to operate a store for about a year and then an inn called the Ukiah House for another year. He then purchased another hotel property at a place called Fountain House between Ukiah and Cloverdale in Mendocino County. He operated both hotels for at least twenty years along with raising livestock. He and Elizabeth had fourteen children, nine of whom were still alive in 1880 (with thirty grandchildren by that time, too). Following an incident in 1868, Harrison was accused of murder. Seemingly straight out of the pages of Wild West lore, the short story is that he had a run-in with a man named John Ketchapaw and in the fray shot and killed him. Fortunately for Harrison, the courts ruled that he had acted in self-defense (Palmer 1880, 332–33, 615–17).

Men came to California for a variety of reasons after the gold rush. Madison E. Mosher was born in about 1846 in Somerset County, Maine. According to the biography, "not taking kindly to farm life," Madison left home at fourteen to seek his way in the world. He fought in the Civil War, receiving crippling injuries. He studied medicine in Cincinnati and returned to Hallowell, Maine, where, after building a successful practice, he fell ill. He was advised to seek a better climate in California, so he moved with his new bride to Salmon Creek, Mendocino County, in 1878 (Palmer 1880, 548–49). At the time the county history was printed, they had one son and were living in the town of Mendocino. However, something happened to his wife Annie, and by 1892 he was remarried to an Alice O., and in 1900 and 1910 he was living in Iroquois County, Illinois. Thus, Madison came to California for his health, but something prompted him to move to the Midwest, once again emphasizing how mobile people could be at this period.

Besides the destination counties studied in this chapter, in northern California there were two centers of initial attraction that supplied a significant supply of settlers to more peripheral locations: the mines and San Francisco. The largely rural areas did not have gold but became quite profitable for many of those fortunate enough to purchase lands they transformed into fertile farms and ranches. That was where lasting wealth was truly to be found in both working the land and servicing those who did. These county founders had great propensity to move, both around the state and back and forth across the United States. Their collective migration histories attest to the fact that California's early statehood was built through the actions of dynamic individuals who knew how to shift and reestablish their residences and careers multiple times.

5

Migrations of the Greenwood, Boggs, and Winkler Families

Traditionally, individuals and families over succeeding generations are responsible for working the land and thus shaping the landscape of particular areas. However, in the frontier and gold rush eras, the familial and spatial relationships among various characters of note are what bring this period to life. Regional history is not simply a mass of swarming people randomly placing their mark on the land; neither is it composed of multiple, unrelated individuals intentionally striking out on their own to create a new place, or more correctly, alter a space in its current state. Indeed, the middle ground that seeks to show how diverse people and families intertwined and acted in concert across the generations lends insight to a view that is more comprehensive than biography, and less overwhelming than more general accounts. Although both the large- and small-scale perspectives are invaluable, the more interrelated approach for this chapter enriches the discussion and is designed to help fit the disparate pieces of this story together.

To catch a glimpse of the complexity of the movements of families over time, I turn to the history of the descendants of three early migrants to California. The first of these was a mountain man and trapper, the second was a former governor of Missouri and early settler of Napa County, and the last was a forty-niner who later became a farmer in Sonoma County north of San Francisco. Their names are Caleb Greenwood, Lilburn W. Boggs, and Clayton Winkler.

The family scale of the migration and settlement model is the focus of this chapter (fig. 1.2). That relationship scale begins small but spreads outward spatially from each particular ancestor starting point. The genealogical connections of these families are paramount in understanding the persistent "stickiness of place" for family members who tended to stay relatively close to the homestead, even through succeeding settlement stages of the surrounding region. The genealogical connections that unify and localize these families through time and space are the primary considerations of this developmental component of California's settlement history.

GREENWOOD AND BOGGS FAMILIES: EARLY CALIFORNIA PIONEERS

As an example of the spatial reach of one family across the California landscape, the case of Caleb Greenwood is very illuminating. He entered the stage before California's statehood, and many of his descendants continued to live in the area for generations. Caleb, known as "Old Greenwood," claimed to have been born in the 1760s, but more likely he was born about 1783 in Botetourt County, Virginia. He first came to California in 1844. Along with other mountain men Elisha Stevens and Isaac Hitchcock, he traveled in the Stevens-Murphy party that made its way from Council Bluffs, Iowa, across the continent and over the snowy Sierra. Caleb was also apparently very involved in blazing a shorter route that went from the Little Sandy River (Wyoming) to the Green River east of Fort Hall (Idaho). This new trail became known as Greenwood's Cutoff, but through some miscommunication its common name became Sublette Cutoff to the many emigrants who later passed that way (Kelly and Morgan 1965, 106–11). The Stevens-Murphy group of pioneers was the first to cross the Sierra with wagons, blazing a trail north of Lake Tahoe (Goetzmann [1966] 2006, 172–74).

The following year in August, Caleb and two of his sons were at Fort Hall convincing 250 to 300 pioneers (about fifty wagons) to follow him to California instead of continuing on to Oregon. Benjamin Franklin Bonney, a young member of this emigrant company described the meeting.

> Greenwood was a very picturesque old man. He was dressed in buckskins and had a long heavy beard and used very picturesque language.... He said the road to Oregon was dangerous on account of the Indians...to California there was an easy grade and crossing the mountains would not be difficult. He said that Captain Sutter would have ten Californians meet the emigrants who would go and that Sutter would supply them with plenty of potatoes, coffee and dried beef. He also said he would help the emigrants over the mountains with their wagons and that to every head of a family who would settle near Sutter's Fort, Captain Sutter would give six sections of land. (Stewart [1962] 1983, 91)

The pioneers who followed Caleb by leaving the Oregon-bound company cemented "Old Greenwood [as] one of California's first civic boosters" (Kelly and Morgan 1965, 149). These California converts traversed the imposing Sierra and continued down safely to Sacramento before winter set in. By the following June, Caleb and his sons had returned to Fort Hall. He even claimed to a traveling companion that he had made at least part of the California to Fort Hall trip four times in two years. It is also possible that Old Greenwood continued traveling east from Fort Hall at this time to collect other members of his family. Regardless, it is noteworthy that Caleb made so many long-distance trips in just a few years (Kelly and Morgan 1965, 181, 189, 205).

Back in California the next winter, Caleb joined in the rescue of the Donner Party as a guide in February 1847. Caleb's son Britton was also part of the second relief party, continuing with the group all the way to Donner Lake, even while Caleb and his son William remained behind in Bear Valley (K. Johnson 1996). Later, on November 2, 1847, Caleb and three of his sons happened to meet Edwin Bryant on the shores of Clear Lake in northern California (Kelly and Morgan 1965, 232, 248–50, 223–24). One of his sons, John Greenwood, and a Calvin C. Griffith were evidently in the area near Clear Lake as early as 1846, so it is not surprising that Caleb would be in the vicinity in 1847 (Palmer 1881, 140–41).

Years before, perhaps as early as 1808, Caleb was a hunter and a trapper on the Missouri River watershed. In approximately 1826, he married a Crow Indian woman (probably half-Crow and half-French), Batchicka Youngcault, and they had eight children while in the Missouri country. The names of two of their children give some clue to the whereabouts and associations of Caleb around the time they were born. William Sublette Greenwood was apparently named after William Sublette, another well-known mountain man whom Caleb knew and with whom he possibly traveled through the West in the company of trappers in 1825 (Kelly and Morgan 1965, 80–81). Governor Boggs Greenwood was named after Lilburn W. Boggs who was elected governor of Missouri in 1836, and Caleb might have known Lilburn when "Boggs was a trader in western Missouri" (Kelly and Morgan 1965, 90).

The mention of the eponymous Governor Boggs is intriguing, because it opens up additional geographic and social connections related to the settling of California. As mentioned previously, Boggs's wagon train traveled much of the way to California with the Donner party, but he providentially separated from them after the Donner group decided to take the misleading Hastings Cutoff (Boggs 1912, 3–4). Migration in America's western frontier had the habit of reemphasizing old associations. In this case, Lilburn W. Boggs had dealings with Mormons both in Missouri and later in California. In the 1830s, the Mormons settling in western Missouri experienced significant conflict with local non-Mormons. Due to the large influx of Mormons and the perceived threat to residents, Governor Lilburn Boggs issued a statement late in 1838 that "the Mormons must be treated as enemies, and must be exterminated or driven from the State" (Marriott 1979, 1). Thereafter, the Mormons were forcibly pushed out of their homes.

Less than ten years later, Lilburn Boggs had settled in California and was busy in his position as alcalde of the Northern District of California. He had dealings with Sam Brannan and other Mormons in the region, but this time their interactions were much more cordial. Boggs even became the distributor in Sonoma County of Brannan's newspaper, the *California Star* (Marriott 1979, 82–83). However, Boggs was still feared by Orrin Porter Rockwell, a Mormon who had been accused (but not convicted) of shooting Boggs in Missouri years before. When Rockwell came from Utah, he participated in mining and running a tavern and inn between Sacramento and Mormon Island known as "half-way house" (Davies and Hansen 2010, 121). "Because he thought that Boggs might take the first opportunity to have him ambushed, Rockwell not only assumed a different name but was constantly armed and had a guard dog always on the alert at his side" (Owens 2004, 294). Fortunately, no blood was shed. But it was reported that Rockwell, under the assumed name of Brown, bested another man in a rifle match at the inn, and in response the defeated man revealed that this Brown was really the Rockwell with connections to Mormon events back East. Drunken citizens of Missouri and Illinois then chased Rockwell out of town (Dewey 1986, 157–58).

Lilburn W. Boggs and his family members were involved or connected with the fur trade and the opening of the West. Lilburn's first wife, Julianna Bent, was sister of southwestern pioneers Charles and William Bent. His second wife was Panthea Boone, granddaughter of Daniel Boone. Lilburn's and Panthea's final years were spent in Napa County, where they are buried. Napa County is the family focal point during that time period. Additionally, two of Lilburn's brothers, Joseph O. and Thomas J., with their families, were early settlers in Sonoma County (next to Napa County), so the Boggs family was well represented in the region.

Most of Lilburn's children settled or spent time in California. These offspring include Angus, Henry C., Thomas, William, Albert, John, Theodore, Minerva, George W., Sophia, and Joseph O., who were all born in Missouri (Family Tree in FamilySearch.org). Only Martha, who died in 1843 in Missouri, did not make it to California. Angus settled in Sonoma County on a ranch that he bought in 1859 in partnership with another western explorer, John Hatcher (Hafen 1972, v. IV, 133). Henry came to California in 1850, first settling in Napa County. In 1864, he moved to Lake County, becoming a banker in Lakeport (Palmer 1881 [1995], 216). Thomas was a friend of Kit Carson and in 1853 accompanied him on an expedition driving sheep purchased from the Navajos to California to sell. He later married a stepdaughter of Charles Bent and helped settle Boggsville, Colorado (now Las Animas; Estergreen 1962, 209; Lavender 1954, 391–92). He passed away in Clayton, New Mexico, in 1894.

William Boggs settled in California with his father and tried his hand unsuccessfully at gold mining at Mormon Island (Boggs 1912; Marriott 1979, 88–89). Albert Boggs lived in Napa County for many years and passed away there in 1897. John Boggs moved to Colfax County, New Mexico, by 1870, working in mining that census year, and later settled in Colorado, dying in Bent County in 1913. Theodore Boggs also left California to seek mining wealth outside the state in 1862. He founded the mining town of Big Bug, Arizona (near Prescott), which was quite a successful mining area for a time, though it is now a ghost town (Sherman and Sherman 1969, 13–14). He died in Prescott in 1905. Minerva Boggs died in Napa County in 1864, after only being married three years. George W. Boggs was married in Sonoma County and later moved with his family to the Stockton area of the Central Valley. Finally, Sophia Boggs lived for decades in Napa County, raising a very large family (she died in San Francisco in 1917), and Joseph O. Boggs lived for many years in Napa County until his death in 1879. Thus, the California ties were strong for the first generation of Lilburns, even if some family members headed elsewhere.

The various human threads in the saga of California settlement thus have come together. California's early population consisted of a rich mix of migrants from all over the United States, but there was a concentration of settlers from the vicinity around the old western edge of the United States (e.g., Missouri, Illinois, Iowa, and Nebraska). Caleb Greenwood and Lilburn Boggs being from Missouri, and the Mormons having their run-in with Boggs in Missouri, all found themselves in a new world where the changed political and settlement environments and opportunities for economic advancement helped assuage some of the old tensions. On the other hand, it is significant that many of these people, such as the mountain men, kept up their associations in the West even after their "glory days." The descendants of all these pioneers continued to make their mark in the new spatial arrangement that was California, while their previous relationships in former spheres in the plains and mountains to the east slowly faded away.

To illustrate this last point, we return to the Greenwoods, where we see the family's outward geohistorical influence. Batchicka Greenwood, Caleb's wife, died while in Missouri, but the rest of the family ended up in California (minus a son named Joseph who is lost to history). In the early gold rush

FIGURE 5.1. Town historical marker and fire bell, Greenwood, California. Photographed with Lori Otterstrom, great-great-great-granddaughter of Caleb Greenwood. Courtesy of the author.

period, the Greenwoods were living in the mining area of El Dorado County. As noted in chapter 2, the town of Greenwood or Greenwood Valley in El Dorado County, located between the South and Middle Forks of the American River, was named after Caleb or his son John (fig. 5.1; Davies and Hansen 2010, 178). Caleb's young eight-year-old daughter, Angeline, died in Coloma about 1849, and Caleb followed in death about 1850 in the same vicinity.

Is it possible that one of his sons was also separately involved in western exploration? In August 1845, another Greenwood went with Lieutenant James W. Abert from Bent's Fort down the Canadian and Arkansas rivers, as directed by John Fremont (Abert 1999, 18; Goetzmann [1966] 2006, 251). Goetzmann thought that this was Caleb. However, here is a case of mistaken identity, because Caleb Greenwood was leading a wagon train from Fort Hall to California with his sons John and Britton in the summer of 1845 (Kelly and Morgan 1965, 149). Abert reported only that a Mr. Greenwood and Mr. Hatcher helped lead the expedition, so it is nearly certain that this Mr. Greenwood was someone other than Caleb (Kelly and Morgan 1965, 320–21). Perhaps it was the mysterious Joseph Greenwood, whose existence is still questioned, but who may have been a son of Caleb. This is a possibility, because John Hatcher and Caleb Greenwood are both presumably from Botetourt County in Virginia (Hafen 1972, v. IV, 125; Thomas 2000). That is interesting because it would connect the Greenwoods with the Boggses via John Hatcher, who presumably interacted with both families. In this formative period of the West, the land was large, but the number of actors was relatively small.

What proceeded in the family migration history may have been repeated frequently by others who first spent time in the Sierra mining communities. All of the Greenwood brothers left the Sierra within the next several years. Indeed, as analyzed earlier, the vast majority of early miners moved through the various mining districts very quickly, so the Greenwoods were following the general pattern. Thus Caleb's sons, James Case, Britton, William Sublette, and Governor Boggs went to Mendocino County in 1854 and settled along the coast in Big River Township, and the community that developed there was called Greenwood. The name was later changed to Elk to lessen confusion with the Greenwood town in El Dorado County (Palmer 1880, 233). While Britton stayed in the Greenwood area of Mendocino County and raised a large family, William Sublette Greenwood and his wife settled in Humboldt County along with his brother James Case and his family. Governor Boggs Greenwood apparently died somewhere near the Oregon-Idaho border before 1870. The other brother, John, was reported to have been living in Marin County until he died in 1859 much farther south in San Bernardino.

Tracking some of the succeeding generations is difficult. John Greenwood may have had one child in Marin County, but the details are inconclusive (Kelly and Morgan 1965, 311–12). Britton had at least nine children while living in Greenwood (Elk), Mendocino County, and there are records of two of them marrying later. Sarah Murphy Foster from the Donner Party "spent the last years of her life in Mendocino County, where she was acquainted with Brit Greenwood's family. She gave them a damaged book that had been salvaged from Donner Lake—damaged because in their extremity, the Murphys had eaten its leather covers" (K. Johnson 2009). Governor Boggs Greenwood was apparently unmarried when he met his untimely end (possibly hung for murdering a man). William Sublette Greenwood died in Humboldt County at the hands of a hit man, having fathered one daughter (Kelly and Morgan 1965, 317–18). James Case had five children while living in the vicinity of Arcata, Humboldt County. Sarah Mojave Greenwood had eight children, one in Anderson

FIGURE 5.2. Caleb Greenwood family locations. Data sources: Kelly and Morgan 1965; genealogical records in possession of the author. Roman Huerta, ThinkSpatial, BYU Geography.

Valley, five in Boonville (both in Mendocino County), and two in Pine Ridge, Fresno County (fig. 5.2). Thus, the second generation from Caleb Greenwood spread across parts of northern California. This relative persistence contrasts with the earlier, cross-continent moves of Caleb and his sons.

A third family example emphasizes how the second and third generations of migrants spread out from initial settling points. These striking patterns are illustrated using the Winkler family that settled in Sonoma County.

WINKLER FAMILY

Clayton Winkler was born January 12, 1831, near Irvine in Estill County, Kentucky. He was one of seven children of David Winkler and Sarah Asbell who were both originally from North Carolina. The family moved from Kentucky to St. Joseph, Missouri, after 1836. There, both David and Sarah died of typhoid in 1840. One of Clayton's younger brothers, named Simpson, also passed away in 1840, being just about four years old. Clayton came west in 1849, with his brothers Archibald, Howard, and David (Jr.), as part of the great gold rush. His younger brother Howard died somewhere along the route to California. The three remaining brothers arrived in Sacramento on October 18, 1849. Clayton spent time in the gold districts, trying to make a profit by finding enough of the elusive metal to save a little gold dust after his expenses. His older brother, Archibald, died in 1851 in Coloma, El Dorado County (the site of Marshall's original gold find in 1848 and where Caleb Greenwood died in 1850). Clayton's only sister, who took care of Clayton and Howard after they became orphans in 1840, never came west but instead married in Missouri and raised a family in Indiana and Iowa. His older brother Sidney also settled far from California. So only David and Clayton were in California after 1851. However, David had two children there and then migrated north to Oregon. Therefore, Clayton was the one persistent Californian of the family.

Experiencing a period of poor luck in the mines, Clayton tossed a coin with a sailor friend to decide whether to stay in the mines or go to sea. The flip of the Spanish *real* directed Clayton and his companion to San Francisco and the Pacific Ocean. However, once there, his stint as a sailor was very short and unpleasant, for he got sick pumping bilge water and never went to sea. By some good fortune, on his return to the mines in 1851, he traveled through Santa Rosa in Sonoma County, where he met the Lancaster Clyman family in Graton. They befriended him and took care of him for a time. He liked the area so much that the next year he bought 280 acres from Mr. Clyman in Green Valley, west of Graton. Clayton's farm was located on some of the most fertile land in California, and that was a key to his later agricultural success. Indeed, Green Valley was the site of the first commercial orchards in Sonoma County beginning in about 1853 (H. Hansen and Miller 1962, 55). He later spent more time in the gold fields of California, as well as working as a wheelwright in Placerville (El Dorado County) for the Studebaker brothers and farming potatoes on his own land. He built a small house on his property in 1860 and then went to the Comstock Lode in Nevada for more mining adventure during 1862 and 1863.

Upon returning to Green Valley, Clayton continued farming and started an apple orchard. In 1866, he married Martha Brain. Martha was a daughter of English parents (Samuel and Sarah Brain) who had migrated to America sometime between 1846 and 1849. Martha was born in 1849 in Pennsylvania. Later, in the mid-1850s the Brain family came to California to the gold districts of Nevada and Placer counties. By 1865, the family had moved to Green Valley (Thomas 1988, 106–12).

Clayton Winkler, Samuel Brain (father of Martha Brain), and thousands of others followed a similar path: gold lured them to the Golden State, but mining was just a temporary attraction that ended after a relatively short period. Samuel Brain presumably mined in the vicinity of Michigan Bluffs, Placer County, where he lived with his wife and four children during the 1860 census. Michigan Bluffs' best years were before 1860, after which the amount of gold taken from the area dropped from $100,000 a month in 1858 to less than $25,000 monthly in 1867 (Gudde 1975, 214). Interestingly, the man who really struck it rich in Michigan Bluffs was none other than Leland Stanford, who made the "nucleus of his fortune" during the early 1850s selling goods there (Gudde 1975, 215). In the case of Samuel Brain, mining only lasted a few more years, and then opportunities in agriculture and commerce to the west of the Sierra lured him to Sonoma County where his daughter settled down with Clayton Winkler. The Winklers had nine children between 1867 and 1889, three girls and six boys, all born in Green Valley (fig. 5.3).

Agriculture was very important to the county at this time. The *Santa Rosa Republican* on December 17, 1891, described the county this way:

> Of the total of 960,000 acres, 200,000 are valley land, the very richest soil, being a black loam; 200,000 acres are rolling or higher table-land with an exceedingly rich soil and considerable sand, the best fruit land; 200,000 acres of foothill lands are adapted to many kinds of agricultural products and pasturage; 100,000 acres of mountain land are adapted to grazing; and about 80,000 acres are of redwood timber lands of the most magnificent growth to be found anywhere. (H. Hansen and Miller 1962, 55)

Of Clayton's six sons, at least three were employed in agriculture in Sonoma County, showing persistence on the land in both location and occupation. Arthur S. Winkler was engaged in different ranching and farming enterprises for short periods until buying the 38.75-acre Surryhue ranch in 1901, eight miles north of Sebastopol. He first planted grapes there. "However, he later changed his plans and put in apples planting the Gravenstein variety"

FIGURE 5.3. Clayton and Martha Brain Winkler family portrait with their nine children ca. 1895. Courtesy of Jean Winkler Thomas.

(*History of Sonoma County* 1926, v. 2, 258). Another son, Edward C. Winkler farmed on 32.5 acres of his father's homestead, growing Gravenstein apples, pears, cherries, and grapes. He was even a member of the board of directors of the Gravenstein Apple Growers Association of Sonoma County (*History of Sonoma County* 1926, v. 2, 321). Walter S. Winkler gained 32 acres of his father's land in the early 1900s, and he too was involved in growing various crops.

Seven of Clayton's and Martha Winkler's children died in Sonoma County, while one daughter, Harriet, passed away in Albuquerque of tuberculosis about a month after moving there in an attempt to improve her health (Thomas 1988, 174). Another daughter, Florence, died in San Francisco at a Christian Science sanitarium (Thomas 1988, 303). Thus, the two children who died away from Sonoma County had left for health reasons. The following generations slowly spread outward from this family hearth. Of Clayton and Martha's twenty-two grandchildren, twenty were born in Sonoma County, and fifteen also died in the same county (fig. 5.4).

The next generation of descendants, the great-grandchildren, were born in close proximity as well. This time some 79 percent (31 of 39) were born in Sonoma County with the remaining eight being born in other parts of northern California. It was not until the generation of the great-great-grandchildren that substantial spread outside of Sonoma County occurred, with 57 percent

FIGURE 5.4. Birthplaces of Clayton and Martha Winkler descendants in northern California from the second generation (children of Clayton and Martha) to the fifth generation (great-great-grandchildren). Clayton and Martha are the first generation and were not born in California. Data sources: Thomas 1988; and genealogical records in possession of the author. Jeffrey Stark, ThinkSpatial, BYU Geography.

TABLE 5.1. Birthplaces of Clayton and Martha Winkler's descendants by generation.

	TOTAL #	SONOMA	%	OTHER CA	%	OTHER USA	%	OTHER	NO DATA	YEARS
Children	9	9	100	0	0	0	0			1867–1889
Grandchildren	22	20	91	2	9	0	0			1891–1920
4th generation	39	31	79	8	21	0	0			1917–1954
5th generation	105	60	57	40	38	5	5			1942–1983
6th generation	122	47	38	36	30	33	27	4	2	1963–2006

TABLE 5.2. Known descendants of the children of Clayton and Martha Winkler through 2006 (includes spouses).

WINKLER CHILD	DESCENDANTS
Sarah Jane	119
Arthur Stephen	14
George Howard	No children
Harriet Lorena	16
Oliver Milton	234
Edward Clayton	32
Samuel Ernest (Ernie)	28
Walter Scott	34
Florence Estelle	14

being born in Sonoma County and 38 percent in other parts of California. This was the first generation where descendants were born in another state (five of the 105 great-great-grandchildren). Although these numbers are less complete than the others, the great-great-great-grandchildren were born in more dispersed locations with only 38 percent in Sonoma County and 30 percent in other parts of California. That left thirty-three who were born in other parts of the United States and four born internationally (two others had unknown birth locations; table 5.1). Thus, the geographic spread of the descendant branches of Clayton's and Martha's family tree illustrates a variety of interrelated trajectories focusing on California and the West.

From the nine children of Clayton and Martha, the multiplication of descendants occurred unevenly. The number of children and grandchildren through all generations up until the 1980s (including spouses) from the issue of Clayton and Martha Winkler, ranged from zero to over two hundred (table 5.2). An important factor in this numerical growth is the number of children from the original nine offspring. Oliver, with the most descendants, had five sons; Sarah Jane, with the second-largest posterity, had six children who grew to adulthood. The rest of Clayton's children had one or two children each (except for George, who died of tuberculosis before having any children). Their descendants up to the 1980s numbered under forty each.

Persistence then, appears to have been the pattern of this family, after Clayton's high mobility before his marriage in 1866. Even with the potential push out of Sonoma County due to marriage of the Winkler descendants, the tendency to stay nearby is remarkable. Part of the stickiness of place regarding this family lies in the fact that the strong regional agricultural economy didn't necessitate relocation. Also, children of Clayton received some of his land after his death (Thomas 1988). Finally, after the rapid settling of the area in the 1850s and 1860s, Sonoma County had consistent economic and population growth that allowed Clayton's descendants to successfully work and live in or near the county of their forebears.

Our three family cases of the Greenwoods, Boggses, and Winklers in California show both the friction of distance and the importance of a family homestead to encourage people to stay put over time. Caleb Greenwood's children were not born in California, and they did not have a particular family gathering place for at least a generation. Caleb's living children, Sarah Mojave, Britton, William Sublette, and James Case, did make their homes in Mendocino and Humboldt counties in the northern part of the state. Lilburn Boggs's family members migrated from Missouri and had conjugal connections with noteworthy frontier families. Once Lilburn settled in the Napa County area, there was a decidedly northern California influence on family locations for years to come. The Winkler's core area of Sonoma County was also well documented. In these examples, the multigenerational migration that occurred was mostly short range, especially into the first half of the twentieth century.

One of Ravenstein's laws of migration concerning short-distance migration appears to hold in these case studies, although crafted for a different population in a different part of the world. Thus, after the initial long-distance migration to California, the Greenwood, Boggs, and Winkler migrants moved relatively short distances.

The emergence of California can be seen as an amalgamation of countless similar stories of immigrant families and individuals who brought with them new energy coupled with longstanding associations that evolved over time. For these examples, spatial proximity of family members was surprisingly stable once the extending generations took root. Chapter 7 will explore this concept of stability at the community level in the Sierra and will illustrate how people tended to settle and remain in place over just a few generations. The stories of the Greenwoods, Winklers, and Boggses attach real names, faces, and places to the early explorers, miners, and farmers of California. They also allow us to get a glimpse of what happened to these families after the initial excitement of settlement gave way to established life in the growing Golden State.

6

Mormons from Yerba Buena to Sutter's Mill

Families such as the Boggses, Winklers, and Greenwoods have made their mark on California from the first family migrants to the present. Even though the families dispersed over time, there is a noteworthy tendency for a continuing descendant presence in those counties that were early migrant destinations. Thus, the migration rate at the family scale is much lower than at the individual scale of the first miners and migrants who came to the Sierra and soon moved on. Our next scale enlargement is to the social network level, illustrated with a peculiar religious group case. The group came to California in the 1840s and made a significant impact initially, only to be later called in large numbers to migrate eastward. This group comprised members of the Church of Jesus Christ of Latter-day Saints, or Mormons.

As introduced in previous chapters, the Mormons played a fascinating part in the early history of California, especially in the gold fields of the Sierra, the San Francisco Bay Area, and in San Bernardino. Their early establishment in the West allowed them to profit from the gold rush via two main avenues. First, they made hefty profits because they were in on some of the early gold discoveries and prospecting in the foothills of the Sierra. Second, Mormons provided goods and services to the miners who anxiously swarmed the gold fields, both in northern California and along the California Trail in the Salt Lake area. As they joined the rapid development, propelled by the golden catalyst, they also played a key role in the growth of a number of communities. The Mormon influx came at a critical juncture in the area's history, when California's population was relatively small in comparison with the flood of migrants that followed on their heels. The fortunate timing magnified the Mormon impact. Additionally, they were a more cohesive group than most, if not all, other gatherings of early immigrants, due to their strong religious ties. This produced a super-interrelated chain of spatial impacts on the landscape.

Two main groups of Mormons made important contributions to the development of northern California during the key early period before and soon after it became a state—namely the ship *Brooklyn* migrants, and the Mormon Battalion. Additionally, a number of other Mormon migrants found themselves in California before and after 1849. These included the pioneer companies, such as William Brown Ide, who came in 1845 and was a prominent participant of the Bear Flag revolt in Sonoma (Holland 1997, 2–5); the Thomas Rhoads group, who arrived at Sutter's Fort in October 1846; and newcomers from youthful Salt Lake City, who mostly arrived after Marshall's discovery at Sutter's Mill in 1848. (Under Brigham Young, the Latter-day Saints arrived in the Utah's Salt Lake Valley in July 1847.)

SAMUEL BRANNAN AND THE *BROOKLYN* SAINTS

We have already met the other important group of early California Mormons who were led by Samuel Brannan. Their adventurous six-month voyage, chronicled in chapter 2, saw several births and thirteen deaths at sea. Newborns John Atlantic Burr and Georgiana Pacific Robbins received names that unmistakably attached them to the oceans on which they were born. As the first immigrants into the newly American Yerba Buena (San Francisco), these pioneers were not only instrumental in the early development of San Francisco, but they also spread out into other parts of California, despite their failed agricultural community attempt at New Hope (J. Davies 1984, 8). Members of this Mormon group, the Mormon Battalion, and a smattering of overland Mormon immigrants also became involved with the gold rush. Over time, some 95 of the *Brooklyn* "saints" migrated to Utah; 160 remained in California; 20 returned to the eastern United States; 3 moved to Hawaii; and the last 3 went to Arizona (ShipBrooklyn.com 2008).

Samuel Brannan was a key figure in the early gold-rush promotion and profiteering, but in the previous year he had tried to convince Brigham Young to bring the saints to settle in California. On April 26, 1847, Samuel Brannan, Charles Smith (a Mormon scout), and another man left Sutter's Fort to cross the Sierra and intercept Brigham Young's pioneer company as soon as possible. They made impressive time crossing snow-covered Truckee Pass in one day and two hours and then continued east. On their way, they passed the site of the Donner Party tragedy where survivors had so recently been rescued, partly through the financial support of Brannan and other San Francisco Mormons. They also encountered and aided Louis Keseberg, who had been left on his own because of alleged crimes in the Donner camp. Continuing on, they arrived at Fort Hall in early June. There they received correspondence from Brigham Young indicating that the saints were planning to go to Salt Lake and not California. Brannan and his companions hurried onward, hoping to convince Young otherwise. They met the Mormons on the banks of the Green River on June 30. However, Brannan could not change Brigham Young's mind, despite all the pleading and energy he expended both there and later the next month after the saints arrived in the Salt Lake Valley. Sam Brannan left the new Mormon headquarters for California on August 9, 1847, never to return (P. Bailey 1943, 57–68).

In San Francisco, Brannan got back to business. Little known to him, he was soon to achieve great prominence. The *New York Herald* printed news of the gold discovery on August 19, 1848, sparking the great rush to California, but the source was a story written by Dr. Victor Fourgeaud in Brannan's newspaper, the *California Star*, on April 1 at Brannan's request (Bagley 1999, 259). The other famous Brannan incident that excited the locals occurred on May 10. Samuel Brannan had returned from the American River, crossed San Francisco Bay, and then, holding "a bottle of dust in one hand, and swinging his hat with the other, passed along the street shouting 'Gold! Gold! Gold from the American River'" (Bagley 1999, 266; Bancroft 1888, 6:56).

The announcement caused such a stir that San Francisco nearly became a ghost town in the ensuing days, and Samuel Brannan was positioned to profit from the mania. After his gold announcement showmanship in the streets of San Francisco, Brannan stayed in the city "only long enough to buy out every store in town of all the stock they would let him have, and on the thirteenth he was on his way to New Helvetia" (Scott 1944, 222). In New Helvetia, or Sutter's Fort, Brannan and his partner C. C. Smith opened up a store. He also busily set up shop at Mormon Island and Coloma. The stores were immediately successful. "Sam's supply ran out the first day, and although he struggled valiantly to keep supplied, it was impossible" (Scott 1944, 223). Later, he propelled the emergence of Sacramento, whose name he convinced Sutter to change from New Helvetia (Holliday 1999, 77). Unfortunately for Sutter, the land that he owned was bypassed in the initial development of the new city (fig. 6.1; Brands 2002, 273–75).

Brannan quickly amassed profits sufficient to make him one of the wealthiest men in California. He continued to involve himself in a number of ambitious schemes (Bagley 1999, 355; Brands 2002, 276). One of the most famous was his brainchild, the Calistoga Hot Springs resort. "Legend has it that too much brandy fueled Brannan's speech at the dedication of his resort, and he proclaimed, 'This will be the Calistoga of Sarifonia!'—a garbled reference to

FIGURE 6.1. Brannan House, Old Sacramento State Historic Park, Sacramento, California. Samuel Brannan helped establish the city of Sacramento. Courtesy of the author.

the famous New York spa, and a confused echo of Brannan's slogan for his new watering place: 'The Saratoga of the Pacific'" (Bagley 1999, 355). Whether this story is true or not, Sam Brannan was instrumental in the early development of Calistoga, which is north of Napa in the Napa Valley. Purchasing with his wealth over 2,000 acres in the late 1850s in the Calistoga area, Brannan went about developing a fancy resort. He later pushed through development of a railroad line that ran 26 miles up the valley from Napa, completed in 1868. However, this superior transportation connection to the Bay Area did not attract enough resort traffic to cover the costs of the resort, railroad, and Brannan's other land-development schemes in the area. Brannan lost the resort in the early 1870s and died poor and nearly alone in Escondido, near San Diego, in 1889 (Bagley 1999; P. Bailey 1943).

John M. Horner played an important role in the early gold-rush years but is less well known (Horner 1904a; 1904b; 1906). Like Samuel Brannan, he was a Mormon passenger on the ship *Brooklyn*. Originally from New Jersey, he had some farming experience that proved very useful. He first planted wheat in the San Joaquin Valley in 1847 but then moved the same year to the Mission de San Jose (present-day San Jose) to farm. His farming activities were a failure that year. "Although I got no dollars out of it, I did get the experience, which I profited by in after years. I had tested the soil in different places, with several different kinds of farm products, and learned the most suitable season for sowing and planting" (Horner 1904a, 572). The year 1848 was not a success either, but the gold rush changed everything. In the fall of 1849, the miners came with huge appetites for fresh produce but found very few local farms to fill their demand. Horner wrote, "As I was the only farmer in the territory who had vegetables for sale, I was much sought after by customers from all sides; two wagons came several hundred miles distant, and bought loads of vegetables at fair prices" (Horner 1904a, 575).

In consequence, John Horner's fortunes rapidly changed, as he became a prime supplier of food for the mining districts and other parts of the new state. In January 1850, his brother William joined him, and they became wealthy farmers in just a few years. Gross sales in 1850 were $150,000, and in 1851 the number increased to $270,000. In the fall of 1851, John was presented with a silver goblet and recognized for his "pioneer" contribution to California agriculture at the state's first agricultural fair by C. A. Shelton, who organized the event (Horner 1904a, 667). His farms and tenants produced 22 million pounds of potatoes in 1853, along with large crops of wheat, barley, and vegetables. Horner stated that "California had not only supplied herself with vegetables this year (1853), for the first time, but she produced a large surplus which could not be sold, and was never sent to market. His rapid business growth continued so that by 1854 John Horner had large land holdings in San Francisco, Santa Clara, and Alameda counties, along with a steamer, a flour mill, a stage line, farmhouses, stables, livestock, tools, and fencing (30 miles in length valued at $900 per mile; Horner 1904a, 670–71). This was the height of his wealth. Over the next several years, he suffered a series of reversals coincident with national economic difficulties. A number of mediocre years followed, and John Horner and family finally left California for Hawaii to raise sugar cane, arriving there on December 25, 1879 (Horner 1904a, 852). So ended John Horner's agricultural rise and fall in the Golden State, with little historical notice.

Even lesser known in the uneven landscape of written history is John Horner's long-time neighbor and one-time friend, Simeon Stivers. He was another *Brooklyn* immigrant and had gone to the gold fields at the onset of the rush with his guardian uncle, Earl Marshall. They did well there and decided to use their earnings to buy a 612-acre farm in what is now Fremont, Alameda. He married Anna Maria Jones in 1858. She had migrated from Utah via San Bernardino with her family. They raised eight children and became a successful dairy and fruit-farming family (Shipbrooklyn.com 2008).

These California Mormons illustrate how such seemingly small decisions on where to live and work had important effects in the development of the state. Additionally, given geographic and religious proximity, their stories are interconnected enough that the stage of action seems to have shrunk to show everyone at once.

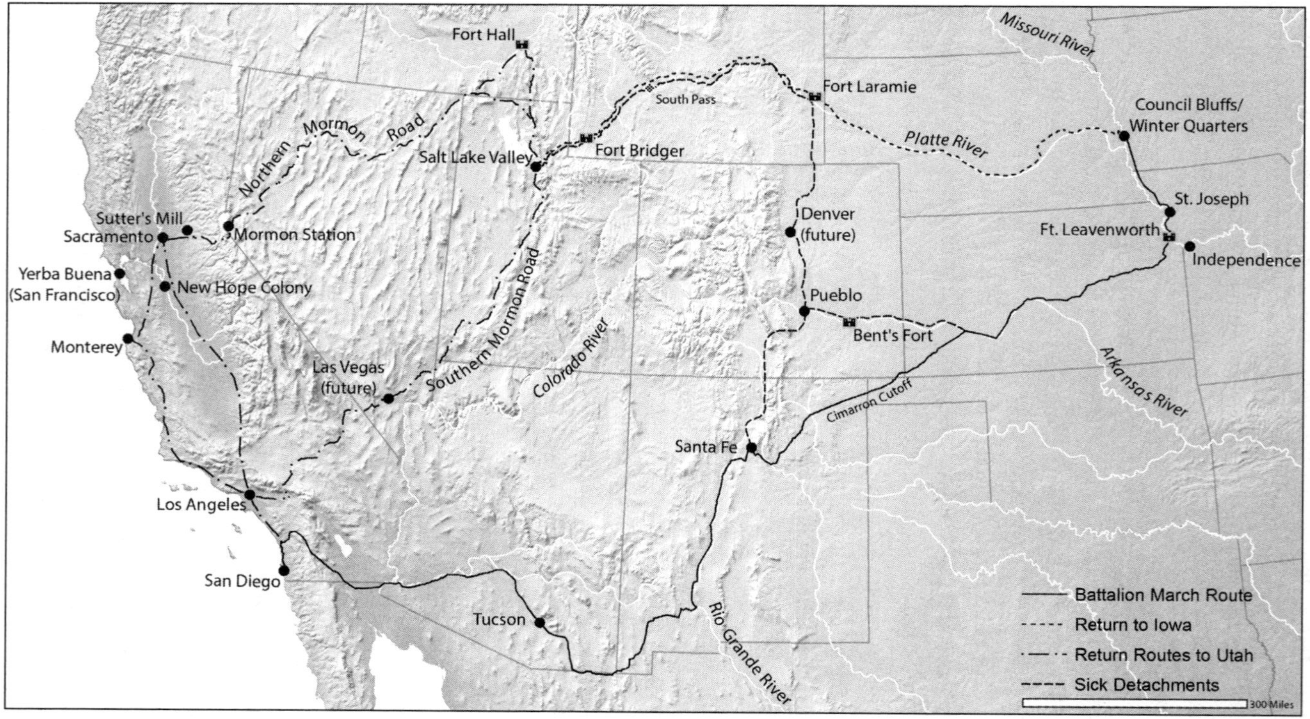

FIGURE 6.2. Mormon Battalion routes 1846–1847. Chris Maderia, Jeffrey Stark, ThinkSpatial, BYU Geography.

THE MORMON BATTALION

In the midst of the Mexican-American War during the summer of 1846, the U.S. government, under authorization by President Polk, recruited over 500 Mormon men to help fight. From among the Latter-day Saints in the areas between Nauvoo, Illinois, and Council Bluffs, Iowa, many joined the cause in July, despite having recently left their homes in Nauvoo under duress. Brigham Young saw an opportunity to help the impoverished Latter-day Saints, as Battalion members would each earn $7 a month, along with an initial clothing allowance of $42 (Owens 2004, 57–58). Some 536 men and a few women began a long march westward in a group that became known as the Mormon Battalion. Along the way, over 150 men and women left the battalion because of sickness and other problems and went to Pueblo, Colorado. About 371 battalion members reached California on January 29, 1847, only to learn that the war was over. Most were released from the army during the summer, and many made their way northward through California in hopes of eventually connecting with the rest of the Latter-day Saints as they came westward from Iowa and Nebraska. But many ended up staying longer in California (fig. 6.2; J. Davies 1984, 5–7). According to Davies, "there were about 250 Mormon Battalion men in the California area as potential participants in the gold discoveries of 1848" (J. Davies 1984, 7).

When James W. Marshall discovered gold at Coloma on January 24, 1848, there were ten people working with him. Five of these were Mormon Battalion veterans. Their names were Henry W. Bigler, James S. Brown, William James Johnston, Alexander Stephens, and Azariah Smith (Bagley 1992; Bigler 1990; J. Davies 1984, 19). Henry Bigler was soon going out quietly during the weekends to seek more gold in other spots along the American River. His finds proved "clearly that the gold in Sutter's Coloma millrace was not an isolated freak of nature" (Owens 2002, 38). After Marshall's initial find, word leaked out, and of course one vigorous propagandist was the opportunistic Sam Brannan, positioning himself to profit from the imminent trade opportunities (J. Davies 1984, 22, 40). The Mormons with Marshall were also soon out prospecting on their down time in the streams and rivers around Coloma. One site they discovered in early March that didn't seem to have much initial promise, became very important. It was on the South Fork of the American River halfway down from Coloma to Sutter's Fort. Sidney Willis and Wilford Hudson, who had been part of the Mormon Battalion, found substantial gold on March 2, 1848 (Cowan and Homer 1996, 109). A month later, gold was discovered in greater quantities in the same area, and not long afterward upwards of 150 Mormons were prospecting in the area (J. Davies 1984, 33). The site became known as Mormon Island.

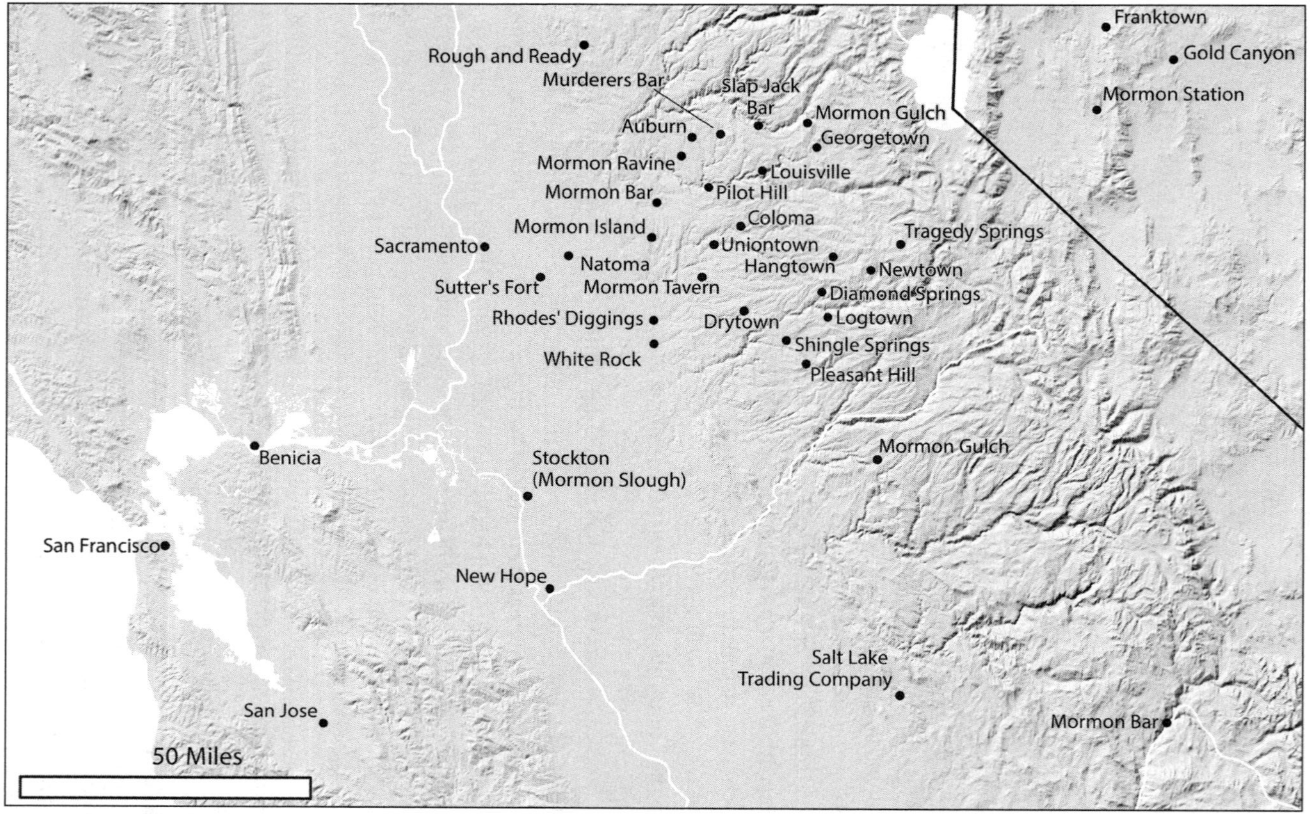

FIGURE 6.3. Early settlements with Mormon connections. Data sources: J. Davies 1984; J. Davies and Hansen 2010; Gudde 1975; Ryan Shields, ThinkSpatial, BYU Geography.

The swirling excitement among a select few increased substantially once the truth of the finds was verified later in the year. The Mormons were some of the earliest miners in California, but their small numbers were soon overwhelmed by other new Californians answering the siren song of gold during the spring and summer of 1848 (Holliday 1999). The next year, many more thousands came by land and sea, and the story of the Mormons in California was nearly swamped in a sea of tales of people from around the world (J. Davies 1984, 393–94). But during the early period from 1848 to 1859, the former U.S. Army soldiers established numerous other mining camps besides Mormon Island, including Mormon Bar, Mormon Gulch, Rhodes Diggings, and New Town (fig.6.3; J. Davies 1984, 355–80).

CONNECTIONS TO UTAH AND ELSEWHERE

Brigham Young, the leader of the Mormons, publicly discouraged Church members from a wholesale rush to the California gold fields. However, not long after remnants of the Mormon Battalion forged a trail east over the Sierra and made their way to the Salt Lake Valley, Brigham Young quietly organized groups of men to go (or to return, in the case of some Mormon Battalion veterans) to California to prospect for gold for the benefit of the Church and its leaders and to organize the Church for the Mormons who were in the foothills, San Francisco, and other parts of northern California. Mormon apostles Charles C. Rich and Amasa Lyman were given specific responsibilities regarding oversight of the ecclesiastical situation in California (Owens 2004, 226).

Thus, a complicated picture of migration among the early Mormons emerged in 1848 and 1849. People were going back and forth between the Utah Territory and California in large numbers in a revolving pattern that most likely differed from the mass of forty-niners. Studying the Mormon pattern of migration and settlement development in California can show the differences between their spatial settlement and movement behavior and that of non-Mormons during the gold-rush period.

What about the Mormon miners and other Mormons who migrated to other areas of the country from California? These people are exceptionally hard to track if they did not stay long in California. Some of them left the region because of failure in the mines or other economic enterprises, while the Mormons who had struck gold were encouraged to go on to Utah, and sometimes to later return to California (Owens 2004). Taking Mormon gold-rush-era settlers identified by J. Davies (1984) from the 1850 U.S. census and other sources to have been in California at that time, a focused search was conducted for these same individuals in 1860 (and later censuses for some individuals)

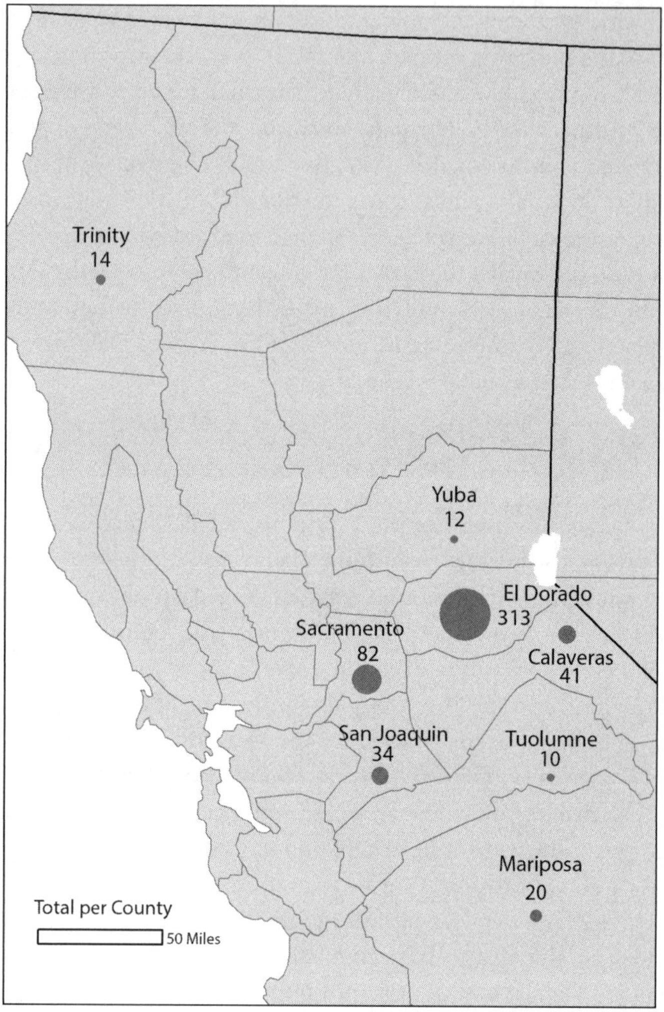

FIGURE 6.4. Mormons in California, 1850. Roman Huerta, ThinkSpatial, BYU Geography. Data source: J. Davies, 1984.

TABLE 6.1. Locations of probable 1850 California Mormons in the 1860 U.S. Census.

LOCATION IN 1860	NUMBER	%
Not found 1860	340	56.7
Dead by 1860	3	0.5
Found Total	257	42.8
Found in California	143	23.8
Found in Utah	40	6.7
Found Elsewhere	74	12.3
Total	600	100%

using census indices from Ancestry.com to identify subsequent places of residence. The exercise quickly revealed that these people remained difficult to track. This was the case even though Mormons were less likely to leave the West than regular forty-niners. Indeed, only about 257 of the nearly 600 1850 California Mormons could be located with some confidence in the 1860 records. Of those found, over half were living in California, 40 were in Utah and 74 were elsewhere in the United States (fig. 6.4; table 6.1).

An example from the Mormon Battalion illustrates that some people did not find California very lucrative or enjoyable, but neither did Utah become the final destination. William R. Tubbs was born in 1824 in Cuyahoga County, Ohio. Sophia Smith was born in 1829 in Orleans County, New York. When the Mormon Battalion was being gathered in Council Bluffs, Iowa, on July 16, 1846, William enlisted in Company D. Sophia Smith also joined as a laundress in Company D. Both of them continued with the Battalion until Santa Fe and the Rio Grande River Valley. There, they were assigned to the sick detachment led by Captain William Wesley Willis (a fourth great-grandfather of the author) and traveled north to Fort Pueblo, and staying through the winter of 1846 and 1847. They continued on to the Salt Lake Valley, arriving not many days after Brigham Young on July 29, 1847. Sophia and William must have become close on their long travels together, as the two were married on August 17 by Orson Pratt (a Mormon apostle). They lived in Salt Lake City until 1849 and then moved to Sacramento, where they stayed from 1849 to 1851. California was not their final destination, but unlike many of the Mormons from the ship *Brooklyn* and Mormon Battalion ranks who migrated to Utah from 1848 onward, the Tubbses first moved from California to Bentonsport, Iowa; then to Racine, Wisconsin, in 1854; and finally to High Forest, Minnesota (near Rochester), where William and Sophia raised a large family and he worked as a farmer. William died in 1897; Sophia lasted several more years until 1904. What was the draw of Minnesota? It is probable that the Tubbses became disassociated from the Mormon faith and that they had family or an opportunity in Minnesota. Once again, the long-distance migration of the Tubbses during the 1840s and 1850s illustrates how actively some people moved, even before the addition of railroad routes and then automobiles.

CALIFORNIA SETTLERS

Aside from the important impacts of the *Brooklyn* immigrants of 1846 and the Mormon Battalion veterans of 1847, a number of other Mormons were involved in the early development of California. Seth and Priscilla Lincoln make a peculiar example. They were not the typical Mormons in early California. Instead, Seth Smith Lincoln found his way into California's Mormon milieu by a completely different route. Seth and his wife Priscilla Lincoln were both from Plymouth County, Massachusetts, born in 1818 and 1819 respectively. They came into the world within a day or two's journey from the sea, and something drew them to adventures far from Plymouth Bay or Buzzard's Bay. For by the early 1840s, they were in the Society Islands (now French Polynesia) with a number of other Americans. What brought them there is unclear, but it is probable that Seth was some type of merchant or trader. Regardless, on August 10, 1844, they were the first people baptized by Mormon missionaries in Papeete (Pratt 1990, 261). They left the Society Islands in 1847 and landed in San Francisco on the ship *Francisco* on June 10 (Pratt 1990, 328). There, the Lincolns became associated with Sam Brannan and other migrants who had arrived the year earlier on the *Brooklyn*. They made their home in San Francisco, and Seth even built a home for the famous Brannan in 1847 (Bagley 1999, 405).

By April 1848, Seth was a leader of the Church congregation in San Francisco, before many of the members removed to the newly discovered gold fields. Addison Pratt, a Mormon missionary in the Society Islands when the Lincolns lived there, visited them in San Francisco in February 1850 (Pratt 1990, 429). Because the 1850 census is missing for San Francisco, it is highly likely that the Lincolns were still there when the census was taken, as they do not seem to be anywhere else in the country. In January 1853, Seth died at sea somewhere near San Francisco (Family Tree in FamilySearch.org), and Priscilla was left a widow. It is likely that the family was living in Massachusetts at the time of his death because Priscilla gave birth to their third and fourth sons, Jasper S. and George S. Lincoln in 1851 and 1852 to 1853 in Massachusetts (Family Tree in FamilySearch.org). Soon Priscilla and her children were back in California. She apparently married a Mr. Sargent, and together they had one child named Francis D. Sargent in 1855. However, by 1860 she was no longer with Mr. Sargent but was living alone with her four sons in Salt Lake City (Familysearch.org [U.S. Census 1860, 17th Ward, Salt Lake County, Utah Territory]). Sometime after 1862, Priscilla moved back to San Francisco, leaving Salt Lake behind. There she apparently stayed, as did her children and their families. Her sons Granville H. and George S. became merchants and had a store on Polk Street in 1887 (Corran 1887, 744).

George Winner was another *Brooklyn* pioneer whose movements underscore the amount of traveling and migration among the early settlers of California. He and his wife lost their infant daughter Sarah on the voyage to California. Soon after arriving in Yerba Buena, their seventeen-year old daughter,

Elisabeth, received a marriage proposal from a seaman named Bazil. She accepted his invitation even though she was some twenty-four years his junior. When they were married on August 15, 1846, they had the noteworthy claim of being the first non-Hispanic couple married in the new American territory. The Winner family made their home in Yerba Buena and were involved in the early growth of the Mormons in that city. He was associated with Sam Brannan until moral disagreements led to a parting of ways. In October of 1846, George piloted the boat *Comet* up the San Joaquin and Stanislaus rivers to what became the short-lived Mormon New Hope agricultural settlement. According to Richard Bullock, he was the first European river pilot to go up the Stanislaus (Bullock 2009, 3–4).

Although John or Caleb Greenwood, or both, have received credit for founding the town of Greenwood in El Dorado County, some Mormons appear to have had a role in this early town-building effort. A Mormon man by the name of Gates, perhaps Hiram Gates, who had come from Utah, was supposedly connected with the Greenwoods. Additionally, Greenwood town was earlier known as Louisville after the son of Lewis B. Meyers, "first-born son of the township" (Gudde 1975, 145). Lewis Meyers migrated to Utah as a Mormon in 1847 (J. Davies 1984, 361). So Louisville or Lewisville was named after a Mormon, followed by Greenwood, named for a non-Mormon early settler. As mentioned, the combination of non-Mormons and Mormons was also in place at the site of the history-changing event of the discovery of gold at Sutter's Mill in 1848.

Other communities in northern California had a decidedly Mormon imprint. These included Slap Jack Bar on the Middle Fork of the American River, Lathrop's Mormon Tavern on the Placerville to Sacramento Road (J. Davies 1984, 103), which was a stopping place for Mormon Apostles Charles C. Rich and Amasa M. Lyman, and Murderer's Bar in El Dorado County, where our old friend Orrin Porter Rockwell operated the Round Tent Saloon (J. Davies and Hansen 2010, 122). Additional toponymic vestiges of the Latter-day Saints in California include Mormon Bars in Mariposa, Placer, and El Dorado counties, Mormon Gulch, Creek, and Diggings in Tuolumne County, and Mormon Gulch in El Dorado County, as well as the most famous one of Mormon Island in Sacramento County (fig. 6.3).

In sum, the Mormon imprint on the northern California landscape occurred in a compressed period from 1845 for about a decade before tapering off under the overwhelming influx of other populations. Their influence on the land was significant, which in the model is the interface between the functional dimension that lasted about ten years and the social network scale that focused on the Sierra to San Francisco but was tied to more distant Salt Lake (fig. 1.2). A major decline occurred with the Mormon Battalion members emigrating on the Carson Valley Route that they blazed in the summer of 1848. After this time, the Mormon influence in California continued with those such as Brannan and Horner who stayed as merchants, farmers, and miners, along with the Mormon gold seekers who came from Salt Lake City after the beginning of the gold rush.

Yet even though Mormon numbers soon paled in comparison with the overall in-migration, their contributions to the early settlement of California continued. Their migration rates (how often they moved around) were rapid at first, like other early settlers, but the rates differed in an important way. A significant number of Mormons followed the counsel of Brigham Young to migrate to Utah, which was against the tide of new settlers from the East, leaving a much-diffused Mormon body. The stayers were consisted of those disaffected with the Church and some others who desired to remain in California even at the expense of losing a great deal of social capital with the exodus of friends and fellow believers. Because of the shared religious beliefs of this distinctive social group, their geographic impact on California was both focused and highly interconnected. Social ties extend near and far over long periods of time. This was certainly the case with the Mormons in early California.

7

Multigenerational Migration, Town Attractiveness, and Settlement Stability

Now that we have explored the rapid migration of the many individuals who made their way across the continent or around it by sea, arriving in the Sierra hoping for riches, then spreading from there to other valleys and hills in the state, we will move to a larger scale. Being often connected by family and social relationships, it is time to look at the familial migration picture. The tremendous flow of migrants who came in the second half of the nineteenth century was unprecedented in history in its size and velocity. While it is difficult to untangle this mass of humanity that apparently moved so randomly and so often, there are hints that migrants originated disproportionately from certain places. And as the cities and towns developed into a structured urban hierarchy, the patterns of migration followed their own constrained trajectories. Also important is the fact that it was most often families and grouped individuals with interlinked migration paths that settled California and created permanence soon after the arrival of the eager prospectors, many of whom arrived alone.

How do multiple generations influence the settling of a place? The settlement process does not involve the sum of random, individual effects on the landscape, but it is the culmination of years of activities of people who are either related by blood or marriage or at least through a personal connection or association. This translates into an emphasis on genealogical records in order to build an analysis at the community scale of the model, with scores of families combined through several generations. For example, when the Winkler family first settled their farm in Green Valley, Sonoma County, as some of the county's first nonindigenous occupiers, that piece of land's future was immediately enmeshed with the Winklers' future, so long as the property stayed in the family. Not only were there local familial impacts but many of the Winkler children and grandchildren can be found in the vicinity of the family homestead generations later (fig. 5.4). Multiply the Winkler family many times to account for other families in the vicinity, and a village or farming region develops.

These various families work together and interact socially and economically with each other and with outsiders who come to their community. Therefore, small-scale geographies of settlement development are caught up in the multi-generational actions of individuals whose local relationships affect their attachments and economic activities.

Contributing to the study of the region's settlement, a more expansive genealogical dataset can be used to view the longitudinally related structure of the communities of northern California. Using focused searches of data from the Church of Jesus Christ of Latter-day Saints' (or Church) FamilyTree at FamilySearch.org, thousands of records were extracted for individuals residing in California beginning in the late 1840s until approximately 1900. This set of people represents a cross-section of the original settling cohort of California. From the sample, the next generations, linked to the first, illustrate the dispersion of families over time across the landscape of the state. Thus, in this chapter, I show how currently emerging technologies can allow for new depictions and geographic analyses of the long-term migration trends that shaped the fundamental culture and identity of California (revised from Otterstrom and Bunker 2013).

Within the family and genealogy dimension of the model is the cohesive effect of family relationships (fig 1.2). Individuals are tied together in time and space by their biological connections and are thus affected by family choices. They do not necessarily act in an economically rational manner, often because ties to parents, spouses, siblings, and even extended family members can structure, or at the very least influence, their work or career and location upon the land. The interconnection of place and people has a most peculiar manifestation in the imprint of families on particular spots over extended periods. Children and grandchildren are therefore likely to exhibit geographic ties to the family homeland. This is called geographic "stickiness" as illustrated by the Winkler family example in chapter 5.

Until now, there has been a gulf between large-scale historical migration research and personal genealogical searching. This separation was tied to the fact that only fairly small threads of migration history are illustrated by one individual's genealogy. The grander patterns were not apparent. Most attempts to connect separate family trees to discern migration trends have been very modest in scope, but new technologies and database structures are allowing genealogical data to become powerful pointers to past family, settlement, and migration geographies.

Currently, electronic genealogies are available that tie together ever greater numbers of people. These have the potential to eliminate some of the problems

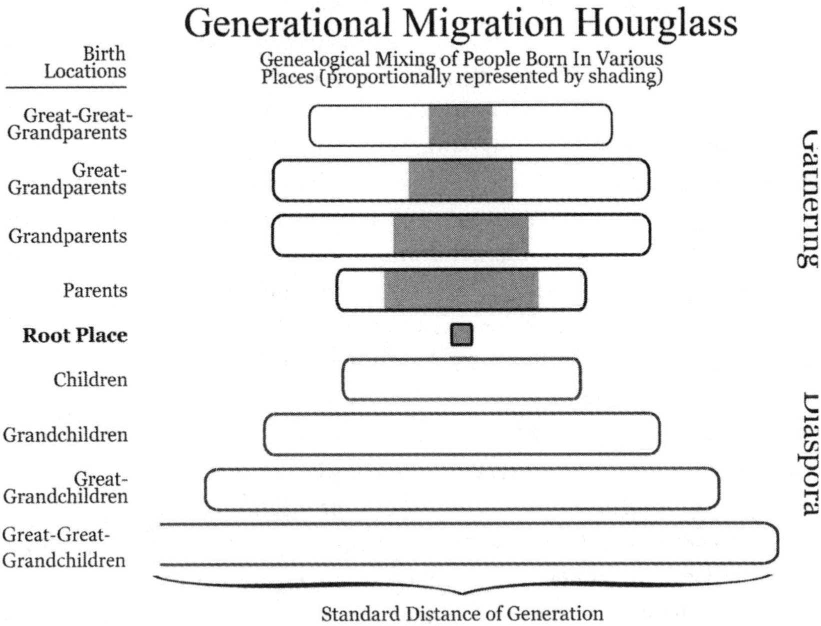

FIGURE 7.1. Generational migration hourglass. Otterstrom and Bunker 2013, 547.

associated with limited multiple-generation indicators in censuses and small-scale place and population compilations of biographies and family genealogies. Migration patterns tied to long swaths of history can be analyzed and illustrated using mapping and statistical measures, illustrating the interconnectivity of places and persons. Additionally, a tool that can be directed to particular times and places while highlighting familial relationships over multiple generations has advantages over previous datasets and analysis methods. Specifically, the FamilySearch online genealogical database has the immense size and robust data structure needed to perform research at various scales and time periods.

PLACE, TIME, AND FAMILIES

Different research threads illustrate the potential of genealogically driven analysis. These four themes are: *gathering*, *genealogical mixing*, *diaspora*, and *community stability*. To clarify, each location (e.g., city, town, village) at a certain time functions as the narrow constriction of the conceptual "generational migration hourglass" (fig. 7.1). Above that narrow opening is the funnel of locations throughout the world from which the people in the constriction were *gathered*, resulting in *genealogical mixing* of ancestral origins. Below is the spreading bottom funnel, which represents the *diaspora* of the succeeding generations. At some places and times, such as during the long periods of agricultural *stability* before the Industrial Revolution, neither the top nor the bottom of the hourglass expand significantly from the point of origin through past or future generations. Migration was mostly local, and it was common for parents and children to live in close proximity (Ravenstein 1976, 182–83).

New World colonization and the Industrial Revolution in Europe changed the migration hourglass (Hoerder 2002; Nugent 1992). In the United States for example, migration from the Old World and subsequent westward movement made the top and bottom portions of the migration hourglass much more expansive than during the earlier centuries of European feudal rural life (Hoerder and Moch 1996; King 2007; Moch 2003). Through succeeding generations, transportation, other technological innovations, and ongoing globalization have redefined the migration hourglasses of communities worldwide. Within this construct, I illustrate how the various hourglasses of combined migration flows define a set of distinct settlement geographies that underscore the central role of families in the evolution of habitable places (Schurer 1991).

In this chapter, I highlight a distinct form of migration here termed "generational migration." Migration analysis over single or multiple generations produces a different set of outcomes from traditional historical migration research because of the distinct time scale involved and the interconnection of the migration streams. Specifically, generational migration shows movement of parents to the birthplaces of children, so it necessarily involves at least three people: the two parents and offspring. Thus, the movement measured for each instance is between two people over a portion of their lives and does not illustrate complete life-course migration (Adams et al. 2002). Many moves during the period between the birth of a parent and the birth of a child could occur without detection by this method, while the common occurrence of a person and children being born in the same town would result in a generational migration distance of zero. However, the ability to follow birthplaces through generations results in a structure that contextualizes other life-course moves in the interim and also displays long-term trends that are completely missed by shorter-term migration studies. Therefore, throughout this chapter "migration" refers to this generational difference of location from the birthplace of a parent to the birthplace of a child. The generational migration hourglass and a wide range of descriptive statistics illuminate the California case study.

Settlement stability reinforces local cultural attributes. On the other hand, migrants build new societies or refashion old ones. In other words, greater numbers of in-migrants increase the potential for cultural change in a community. Generational migrants, then, transfer traditions, languages, religion, and other characteristics from their personal and ancestral places of origin to these continually evolving locales. Thus, a concentrated study of migration over multiple generations intertwines cultural geographies of one place and time with those of another through genealogical ties that can be as thin as a single thread or as thick as a heavy rope, depending on the size of the generational migration.

METHODS: GENERATIONAL MIGRATION EXTRACTION AND ANALYSIS

In this section, I explain how a large genealogical database can be used to reconstruct the multigenerational origins of migrants to various communities at different time periods. I describe a set of genealogical migration measures that illustrate generational migration hourglasses. This type of comprehensive analysis has not been feasible before, and both the questions explored and the methods introduced add new dimensions to historical migration studies.

FamilySearch

This approach extracts straightforward summary measures from the expansive data available in FamilySearch. The FamilySearch Family Tree database is a user-contributed, lineage-linked repository of family history information managed by the Church. There are hundreds of millions of names in the database, including over 155 million from North America. Because authorized users can combine duplicate records, question the veracity of relationships, and add new names that have valid sources, the database is continually improving in quality, which is a step above static published genealogies. In this case, all of the searches were performed within the Family Tree database (databases in other parts of the FamilySearch site, such as the historical country censuses, were not used [Crume 2011]).

Since the accessible database only contains names of deceased individuals, the size and completeness of the dataset decline substantially near the turn of the twentieth century. About 89 percent of the people with life-event dates were born before 1875, with an additional 8.4 percent born between 1875 and 1900, leaving 2.7 percent with birthdates after 1900 (Clarke 2011). Therefore, searches done before 1900 in North America, such as for post-gold-rush northern California, are the most representative of overall historical migration patterns because of the vast number of relevant records in the database.

An online query system developed by Brian Bunker and the author (Otterstrom and Bunker 2013) uses the FamilySearch Family Tree. This method made it possible to sample individuals from a community born during a certain timeframe and then extract their five-generation pedigree information. The five-generation query results were then analyzed and spatial statistics and indices calculated to display on a web page template complete with an option to include maps. In this chapter, the "root" generation individuals are those born in the query place and time, the "first generation" are the parents of the root generation, the "second generation" are the grandparents of the root, the "third generation" are the great-grandparents of the root, and so forth.

Due to search constraints imposed by the FamilySearch application program interface (API), we made adjustments to streamline the searches, including dropping partially relevant records and building a local place database. Most of the individuals in the Family Tree database only have birthplaces and dates listed, which is a limitation. However, the great value of this system is the connection of multiple generations, which allows for long-term generational migration study of families, even with spotty death information. The analysis was done by comparing the birthplaces of children with birthplaces of parents, grandparents, and great-grandparents. By aggregating the root individuals, one can create a cross-section of the migratory patterns of the community. In this way, the conceptual forward and backward hourglass of family relations migration can be illustrated.

Generational Migration Themes and Statistics

As mentioned, this chapter focuses on the community scale of settlement development and migration within northern California that springs from the personal/family scale, beginning with variations in the shape of the personal generational migration hourglass, which is a person's genealogical geography. Second, it expands outward toward the composite hourglass that is created by collating large numbers of records from the same local area, which results in a distinctive community genealogical geography. Again, the conceptual framework revolves around the four main generational migration themes of *diaspora, gathering, genealogical mixing*, and *community stability* using a case study and specialized statistical measures. Each theme underscores important migration processes at the aggregate multiperson level.

Hourglass genealogical geographies are useful in illustrating the geographic persistence or stickiness related to family migration, as well as the unique differences of communities' population stability and migration origin fields. Basic descriptive spatial statistics are necessary to describe the underlying movements of people and to calculate other migration measures. We developed and adapted the following generational migration measures to help quantify and describe the generational migration hourglass for parts of northern California:

- Mean Migration Distance (MMD): average distance between the birthplaces of parents and children who were born in a specific city or county at a certain time;
- Standard Distance of Generation (SD): relative dispersion statistic of all birthplaces from the weighted mean center in one generation;
- Dispersal Index (Ancestor or Descendant, ADI/DCI): mean migration distances compared and computed between generations;
- Concentration Index (Ancestor or Descendant, ACI/DCI): weighted mean center and standard distance of generation compared between generations;
- Genealogical Mixing Ratios (GMR): proportion of ancestors from various source regions/countries;
- Community Stability Index (CSI): ratio of parents born within 40.2 kilometers to parents born farther away;
- Ravenstein Close Migration Index (RCM): ratio of parents born within 80.4 kilometers to all parents in that generation.

Grass Valley and Oakland in northern California are useful illustrations of these statistics, because they varied greatly in size, history, and relative location (table 7.1; fig. 7.2).

For individuals born in 1900, the mean migration distances (MMD, root–1st) were similar between the cities for the first generation (2,166–2,307 km). However, the grandparents (2nd generation) in the two cities had greatly different migration distances from their place of birth to their children's (2,445–1,283 kilometers). Standard distance is the square root of the averaged squared distance of multiple points to a central location. It is not the same as the aspatial standard deviation, because all differences from the mean are necessarily positive (Rogerson 2010, 38). Thus, the standard distance of generation (SD) is a useful descriptive measure that quantifies the relative distribution of one generation's birthplaces. A large SD points to dispersed birth locations and an expansive source area for that generation's migration; while a small SD value could indicate chain migration effects because of a concentration of nearby birth locations of parents who may have migrated in tandem. To differentiate the two measures: the MMD is the average generational migration distance of all *parents to their children*, while the SD describes the spatial arrangement of all the people within *one* generation. The other indices that add additional explanatory power between generations are explained below.

TABLE 7.1. Example statistics from migration genealogy results obtained from FamilySearch (in km).

STATISTIC	GRASS VALLEY, NEVADA CO., CALIFORNIA, 1900	OAKLAND, ALAMEDA CO., CALIFORNIA, 1900	EXPLANATION
# People born in place during year (root generation)	273	434	
# People in 1st generation	434	689	Number of records extracted in each generation
# People in 2nd generation	550	887	
# People in 3rd generation	616	953	
# People in 4th generation	618	1153	
MMD root–1st	2,165.5	2,307.5	Mean Migration Distance
MMD 1st–2nd	2,445.1	1,282.5	
MMD 2nd–3rd	667.1	333.1	
MMD 3rd–4th	99.2	190.0	
ADI (1st–2nd)/(root–1st)	1.129	0.556	Ancestor Dispersal Index (km)
ADI (2nd–3rd)/(root–1st)	0.308	0.144	
ADI (3rd–4th)/(root–1st)	0.046	0.082	
MC Root	39.2192 / 121.0600	37.8044 / 122.2697	Mean center of generation (lat./lon.)
MC 1st	39.3435 / 110.9066	39.3492 / 105.3210	
MC 2nd	40.4623 / 91.2182	40.9475 / 85.5956	
MC 3rd	39.9148 / 80.2483	40.6946 / 78.8015	
MC 4th	39.4405 / 77.4138	40.4708 / 76.3166	
Dist. between MC root–1st	875.7	1,484.6	Distance between mean centers of generations
Dist. between MC 1st–2nd	1,684.7	1,686.5	
Dist. between MC 2nd–3rd	935.5	573.7	
Dist. between MC 3rd–4th	248.8	211.8	
SD 1st	3,990.0	3,825.8	Standard distance of generation
SD 2nd	5,313.7	3,792.5	
SD 3rd	4,416.3	3,065.1	
SD 4th	3,716.7	2,552.8	
ACI 2nd/1st	1.332	0.991	Ancestor Concentration Index
ACI 3rd/1st	1.107	0.801	
ACI 4th/1st	0.932	0.667	
CSI 1st	146/219	117/412	Community Stability Index by generation
CSI 2nd	30/369	27/542	
CSI 3rd	5/374	4/567	
CSI 4th	0/326	0/663	
RCM1	152/365	121/529	
RCM2	33/399	28/569	Ravenstein's Close Migration Index by generation
RCM3	5/379	5/571	

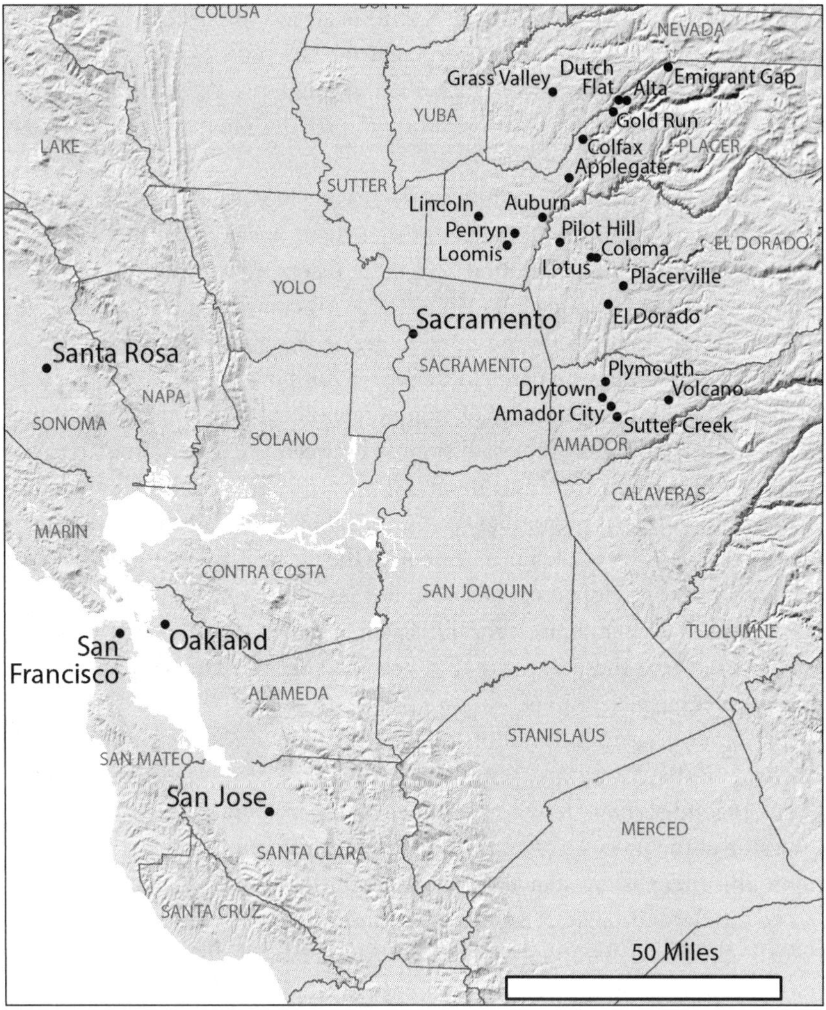

FIGURE 7.2. Northern California study communities. Ryan Shields, ThinkSpatial, BYU Geography.

Diaspora and Gathering

The geographic composition of the gathering represented by the top half of the generational migration hourglass and the diaspora of the bottom half can be described by two related indices: dispersal and concentration. Although each index can be used as a migration description for ancestors or descendants, the indices will both be used in relation to the ancestral theme. The ancestor dispersal index (ADI) is a comparison of MMD among different generations. The MMD between the generations closest in time to the root location (usually between the root and the first generation) is always the denominator in the ratio. These numbers represent the degree to which a person's direct-line ancestors' birth locations spread from each other in comparison with the first generational migration, and they show absolute and relative movement of kin whose families originated in a certain place.

Index values less than 1.0 indicate decreasing MMD, or less long-distance mobility through time, while values greater than 1.0 show that migrations of relatives in successive generations are more dispersed, or more mobile. For example, the ADI for 1900 Grass Valley (table 7.1) comparing the first two generational migrations is 1.129, pointing to greater migration dispersion for the grandparents than for the parents of children born in Grass Valley in 1900. In Oakland, the like value was .556. The grandparents of Oakland-born children in 1900 migrated much shorter distances between the generations than the parents. Explanations of this difference between the two cities are the gold-rush effect (with long migrations) on Grass Valley, and a large uptick in international migration to Oakland at the end of the 1800s. The great-grandparents and great-great-grandparents tied to both these cities had very short MMDs, pointing to a lack of migration for these people and very little dispersion (most born 1820–1840). These statistics thus become a relative indicator of *diaspora* over succeeding generations.

Another measure is the ancestor concentration index (ACI), which indicates the relative localization of ancestors in terms of their origins. In other words, the index shows how concentrated ancestors were in space, and it is an indicator of the propensity of ancestors to be clustered in preceding generations. This index is the standard distance of births of one generation divided by the standard distance of a generation closer in time to the root generation (usually the first generation, or parents). An index value over 1.0 means that the relevant generation (e.g., grandparents or great-grandparents) were more spread out in their birthplaces than the comparison generation (e.g., the parents' generation). This backward-looking ACI reveals the size and proximity of the ancestor pool of people born in one community and illustrates the source from which the people born in one destination place were *gathered*. The relative difference in this index for various times and places allows for comprehensive comparative interpretations.

At first glance, we might expect the values of ACIs to be greater than 1.0, because the second and subsequent generations' starting points would hypothetically be more spread out than those of the first generation. For example, the standard distance for parents of children born in Grass Valley in 1900 was 3,990 kilometers, and for the grandparents it was 5,314 kilometers. Thus, the ACI is 1.332, which corresponds well with the far-flung origins of migrants to

Grass Valley (such as the Cornish). During the growth period of underground mining, population in Grass Valley increased from 3,840 in 1860 to 6,688 in 1880, meaning that at the approximate 1860 birth period of the grandparents many were living elsewhere in the world. The fourth generation in Grass Valley (great-great-grandparents) had an SD of 3,717 kilometers, making the ACI smaller between the first generation and these fourth-generation ancestors (.932). Additionally, Oakland's ACI indicates greater concentration or clustering farther back in time. Instead of the generational migration hourglass spreading outward distance-wise, it was actually contracting in both SD (concentration) and MMD (dispersal).

Community Stability and Local Stickiness

At the aggregate scale, the Community Stability Index (CSI) is a helpful measure. The ratio is a more focused consideration of population gathering or dispersal. This measure can test whether specific settlements tended to be more migratory at different times. It shows how often certain families in an area turn over in population from one generation to the next and focuses on movements between generations. This is the ratio comparing the number of parents who were born in about the same place as the root person (within 40.2 km [25 mi], or about a one-day journey by wagon) with those who were born farther away. The larger the CSI is, the greater the stability in the community. A CSI of 1.0 means that equal numbers of parents were born in an area and elsewhere, with higher numbers indicating greater stability, and values between 0.0 and 1.0 showing greater volatility. Taking a random sample of communities across the nation or other more local geography at specific times can show relative community stability among places. Maps of these patterns can illustrate the regional index differences.

The CSI is useful to illustrate the differential geographic stickiness of various places across the United States and the world at certain historical periods. Specific questions concerning the population stability of towns and counties can then be answered and compared directly with places near or far to generate dynamic matrices and maps of historical long-term mobility. Additionally, these methods permit pinpointing of certain times when single or multiple communities moved from being newly settled to maturely stable, defined by the multigenerational propensities of parents and children to be born in the same place. This helps to reveal migration waves over time as well as settlement stability.

Ernest Ravenstein posited that most migration occurs over short distances (Ravenstein 1976; Grigg 1977). Although his studies focused on Britain in the

1800s, we can test this supposition for other places and periods using Ravenstein's Close Migration Index (RCM) and MMD. The RCM is the same as the CSI, except that the radius is increased from 40.2 to 80.5 kilometers (50 mi), and the denominator is the total number of people in that generation. For example, 41.6 percent of the parents of children born in 1900 were from within 80.5 kilometers of Grass Valley, while it is 21.2 percent for Oakland (table 7.1). Those percentages represent the portion of the samples that moved short distances over the period. Economic gyrations have also precipitated greater mobility and less community stability, and research has explored the associated migration patterns (Hoerder 2002). However, the tie of family relationships over generations strengthens the research related to the inertia of place, or geographic stickiness, even after the advent of easier travel methods.

Community Ethnic Origins

Genealogical identity of a place is computed using the Genealogical Mixing Ratios (GMR) from the distribution of birthplaces for each generation out of the total group of individuals born at a certain time and place. This measure considers the percentage of ancestors from one or a group of countries as one factor and time decay for how many generations the ancestors resided in the United States (taking into account the differences of known versus unknown ancestors) as the second factor.

Calculating and then aggregating the GMR for hundreds born at approximately the same time in one community yields a composite ratio of various regional or national origins for the community at that period. Over time, the same percentages can demonstrate how the identity of a community changes and becomes more or less "American," depending on migration patterns. This is a more far-reaching analysis than can be accomplished with U.S. Censuses, even in the late 1800s and early 1900s, because the identities of grandparents and great-grandparents enter the equation. Additionally, this analysis can be used for earlier periods in American history. Such calculations are valuable to historians, geographers, and genealogists, because they reveal the ethnic character of towns or country subregions in a comprehensive way at many different periods.

Methods Conclusion

Because the generational migration scale is different from that used in other migration studies, let me emphasize the context of migration processes in this study. Settlement migration, intra-urban migration, chain migration, international migration, rural-urban migration, labor migration, and so forth, are all contained within this generational rubric, even though different types of migration have their own particular modes and outcomes (Bailey 2005). It is therefore important to call attention to the possible processes that brought the movement or nonmovement (stability) between generations, even while recognizing that more individually based analyses (such as life-course studies [e.g., Courgeau 1985]) can explain such processes more precisely. Instead, the methods here allow reconstruction of the component migration characteristics of communities spanning many generations that contain various types of migration flows.

The California Gold Rush case study includes an analysis of the first half-century of generational migration in northern California to explore some of the migration themes. It includes a comparison of five commercial cities west of the Sierra Nevada with twenty-three foothill mining communities in terms of their differential attractive pull to migrants in distance. Mining towns are designated as "small towns" and commercial cities as "large cities," per their recent relative sizes. The MMD is used to determine where the longest-range migrants went, the CSI to see whether commercial city dwellers were less migratory than mining city inhabitants, and the RCM to show the relative share of short-distance generational migration.

Stability and Migration Distances: California's Gold Rush and Settlement Creation

After the 1848 Sutter's Mill gold discovery, northern California experienced a rush of new settlement offering a valuable case study of migratory process and settlement development. Certainly, as prospectors arrived, they founded new towns and cities at a dizzying rate. However, many of these places had fleeting lifespans (illustrated in chapter 8), as gold played out and people left nearly as quickly as they came. While some cities faded away, others managed to survive despite the significant volatility caused by in- and out-migration. In each generation of a certain place, the residents are classified as stayers or movers. In some unstable cities or counties, the movers greatly outnumber the stayers. The opposite can be true for other times or places. We used multigenerational migration tracking to compare movers and stayers among different-sized cities and to measure community stability.

The twenty-three smaller communities chosen were in the gold mining counties of Placer, El Dorado, Amador, and Nevada. Those communities are

TABLE 7.2. Mean Migration Distances (MMD) of parents to select California cities, 1850–1900 (in km).

CITY SIZE	1850	1860	1870	1880	1890	1900	AVERAGE
Small	2,890	3,083	3,040	3,058	2,495	1,784	2,725
Large	2,210	2,623	2,466	2,130	2,602	2,084	2,353
Large City							
Oakland	2,090	2,424	1,243	1,283	3,688	2,307	2,173
Sacramento	2,409	2,782	2,747	2,304	2,100	1,534	2,313
San Francisco	2,863	2,834	3,377	3,132	2,759	2,462	2,904
San Jose	1,814	2,517	2,830	1,912	2,262	1,968	2,217
Santa Rosa	1,350	2,508	2,292	2,213	1,887	2,033	2,047
Small Town							
Auburn	3,185	2,575	2,393	2,016	1,690	1,645	2,251
Coloma	3,679	3,152	1,846	902	822	1,718	2,020
Grass Valley	2,793	3,425	3,686	4,362	3,233	2,165	3,277
Placerville	3,256	2,971	2,838	2,351	1,953	1,933	2,550
Sutter Creek	3,096	3,661	2,567	2,592	3,545	2,413	2,979
Volcano	2,404	2,311	2,189	1,380	506	502	1,548

compared with the commercial cities of San Francisco, Sacramento, Santa Rosa, Oakland, and San Jose (fig. 7.2). The database searches were for people born in these cities and towns within five years on either side of 1850, 1860, 1870, 1880, 1890, and 1900. The large query sizes, especially later in the period, give confidence in the validity of the conclusions.

Overall, the mining cities had higher mean generational migration distances. The average MMD from parents' birthplaces to the smaller towns in the Sierra foothills was 2,725 kilometers over the five decades. For the commercial cities, the average figure was 2,353 kilometers. Of the cities, San Francisco had the highest MMD at 2,904 kilometers, while Santa Rosa registered the lowest MMD of 2,047 kilometers over the study period. The other cities were generally between these two figures, except Oakland's MMD rose rapidly at the end of the century and actually exceeded San Francisco in 1900 (table 7.2).

A detailed look by decade provides a more nuanced distinction between the gold-rush towns and the large commercial cities. Between 1850 and 1880, the Sierra cities had larger weighted MMD from the parents to the children born there than the commercial cities to the west; while from 1890 to 1900, the big cities had the advantage. This is explained by the "worldwide contagion" that brought prospectors from across the globe to the Sierra (Holliday 1981, 83–111). The commercial cities also benefited from this tremendous influx from the East and abroad, thus they had large mean generational migration distances as well. The mining towns rapidly declined in MMD from a fairly stable 3,050 kilometer level from 1850 to 1880 down to 1,784 kilometers by 1900; while the large cities fluctuated closer to the average over the five decades, ending above the mining cities in 1900. This analysis illustrates that the gold fields behaved as larger cities in terms of their drawing power for migrants during the gold-rush period: the perception of abundant high-paying jobs in mining could attract people from great distances to even fairly small towns in the Sierra.

After the gold rush, especially post-1880, many mining towns quickly lost their pull, while some others gained strength in their ability to attract long-distance migrants. For example, Placerville's MMD dropped from 3,256 kilometers in 1850 to 1,933 kilometers in 1900; while Auburn's went from 3,185 to 1,645. Coloma declined abruptly in both MMD and migrant numbers, so much so that the 1900 sample only returned three people with four known parents, making the 1,718 MMD very suspect for that year. Volcano also dropped off the chart to about 502 kilometers for MMD in 1900. Two cities in the Sierra foothills that became less provincial in the mid- to late-nineteenth century were Grass Valley and Sutter Creek. One explanation is that the area around these cities successfully developed underground gold mining that continued to draw long-distance migrants into the twentieth century.

Comparing the male and female mean migration distances (distances from birth location to birthplaces of children) for the first generation is not as conclusive. For all large cities averaged for all years, men migrated farther by nearly 80 kilometers (2,404 versus 2,326), while the men and women in smaller cities migrated about the same distance as each other (2,729 for men versus 2,718 kilometers for women). This actually means that the men and women migrated very similar distances within their respective city classifications. Beyond this general observation, the differences between men and women seem to be very city dependent. For example, Santa Rosa had four decades (1870–1900) in which the women migrated much farther on average from their births to Santa Rosa than the men. On the other hand, Oakland and San Jose had men migrate farther in 1870 and 1880.

In measuring the transience within these twenty-three communities, there was a marked difference in community stability (CSI) or population stickiness between and among the different types of cities (table 7.3). In the 1850 period, the large commercial cities showed approximately 23 percent of the children's parents were born within 40.2 kilometers of their respective cities. San Jose (0.521) and Santa Rosa (0.392) had the largest CSI values, as they were fairly small, and many citizens of these towns had been in the area for a while under the Mexican flag. The CSI dropped in half for 1860, as these cities grew very rapidly due to in-migration. The CSI number increased gradually over the ensuing decades until it stood at 0.289 in 1900, which indicates maturing settlement systems, albeit still welcoming substantial immigration.

TABLE 7.3. Community Stability Indices (CSI) for first generation (parents to children) in select northern California cities, 1850–1900.

CITY SIZE	1850	1850	1860	1860	1870	1870	1880	1880	1890	1890	1900	1900
Small	23/291	0.079	121/1093	0.111	124/1155	0.107	205/1131	0.181	352/1074	0.328	357/696	0.513
Large	271/911	0.297	229/1732	0.132	266/1813	0.147	331/1673	0.199	479/2104	0.228	540/1871	0.289
City												
Oakland	37/134	0.276	55/245	0.224	26/203	0.128	42/210	0.200	58/591	0.098	117/412	0.284
Sacramento	40/266	0.150	25/395	0.063	55/422	0.130	89/352	0.253	111/362	0.307	115/338	0.340
San Fran.	36/195	0.185	45/303	0.149	83/369	0.225	75/400	0.188	125/383	0.326	129/374	0.345
San Jose	138/265	0.521	89/381	0.233	65/400	0.163	71/305	0.233	101/397	0.254	100/382	0.262
Santa Rosa	20/51	0.392	15/408	0.037	37/419	0.088	54/406	0.133	84/371	0.226	79/365	0.216
Auburn	5/20	0.250	55/123	0.447	29/110	0.264	23/113	0.204	33/106	0.311	61/104	0.587
Coloma	0/16	0.000	5/35	0.143	4/18	0.222	6/6	1.000	5/10	0.500	0/4	0.000
Grass Valley	12/85	0.141	37/324	0.114	47/402	0.117	52/415	0.125	142/381	0.373	146/219	0.667
Placerville	3/67	0.045	0/222	0.000	15/176	0.085	46/110	0.418	58/136	0.426	34/96	0.354
Sutter Creek	0/5	0.000	1/51	0.020	2/80	0.025	9/79	0.114	20/83	0.241	21/52	0.404
Volcano	0/15	0.000	3/59	0.051	4/36	0.111	0/16	0.000	5/11	0.455	10/12	0.833

(Total # of parents born within 40.2 km of the listed city of their child's birth/Total # of parents born farther than 40.2 km from their child's city)

TABLE 7.4. Ravenstein's Close Migration Index (RCM) for small and large cities of northern California, 1850–1900.

SIZE	1850	1850	1860	1860	1870	1870	1880	1880	1890	1890	1900	1900
First generation (parents to children)												
Small	23/314	7.3	126/1214	10.4	132/1279	10.3	228/1336	17.1	388/1426	27.2	391/1058	37.0
Large	287/1182	24.3	246/1961	12.5	293/2079	14.1	343/2004	17.1	534/2583	20.7	636/2411	26.4
Second generation (grandparents to grandchildren)												
Small							36/1359	2.7	66/1508	4.4	58/1246	4.7
Large							63/1994	3.2	73/2051	3.6	158/2741	5.8

(Total people born within 80.5 km of child or grandchild's city/Total in generation)

On the other hand, the mining towns moved from significant instability to notable stability. These towns were very unstable during the mining peak, with only 7.4 percent of children's parents born within 40.2 kilometers of the respective towns in 1850, explained by the fact that very few settlers lived in the area before 1850. This number increased only slightly over the next two decades (hovering around 10 percent), as thousands of miners flooded in from around the world. However, from 1880 to 1900, the mining town CSI values increased considerably from 0.181 in 1880 to 0.513 in 1900. For example, two bigger Sierra foothill communities of Grass Valley and Auburn had stable CSI values in 1900, 0.667 and 0.587, respectively. Thus, one can give a general estimate of 1890–1900 as the point when these communities made the overall transition from unstable resource extraction centers to stable communities. Additionally, the mining cities were more stable than the larger cities by 1900, while the RCM shows that in both the commercial and mining cities a significant number of parents had moved less than 80.5 kilometers between their birthplace and the birthplace of their children.

Ravenstein argued that most moves are short distance. In terms of generational migration in California in the late 1800s, a high proportion of parents did not move far from their place of birth to where their children were born (table 7.4). The numbers generally increased through the time period: in 1900, 37 percent and 26.4 percent of mining town and commercial city parents, respectively, had been born within 80.5 km. The second generation RCM also reveals that many grandparents had been born near the birthplaces of their future grandchildren, which means that multiple families came to stay in the midst of the gold rush. This fact sometimes gets lost in discussions about the frenzied pace of migrant transience.

Community Origins

A final dimension is the differential origin of the people in northern California. How do genealogical mixing ratios compare between larger commercial centers and mining centers? The four cities of San Francisco, San Jose, Oakland, and Sacramento were contrasted with the four mining towns of Auburn, Jackson, Nevada City, and Placerville for 1850, 1870, and 1890. The most noticeable difference at the height of the gold rush in 1850 is the greater relative numbers of mining-center children whose parents were born in Europe, compared with the other cities to the west (20.5 percent to 10.7 percent; table 7.5). Also, Midwesterners flocked to the gold fields in greater relative numbers (19.4 percent versus 9.2 percent in the larger cities). The international appeal of the gold

TABLE 7.5. Birth origins of parents of children born in select small towns and large cities in northern California, 1850, 1870, and 1890.

SOURCE AREA	1850%	1870%	1890%
Small Towns (1st Generation)			
California	8.5	12.0	31.6
West	0.0	0.2	1.3
South	11.4	5.7	2.5
Midwest	19.4	17.1	11.9
East	10.6	10.9	4.7
New England	5.4	5.5	4.3
Mexico	0.0	0.1	0.1
Canada	0.0	2.6	2.5
Europe	20.5	19.8	17.7
U.S. other	0.2	0.2	0.4
Unknown	23.8	26.0	21.8
Other	0.2	0.0	1.2
Large Cities (1st Generation)			
California	17.5	10.9	26.4
West	0.4	0.3	1.4
South	7.6	6.4	2.4
Midwest	9.2	10.5	11.9
East	11.2	9.4	6.1
New England	11.1	9.1	4.3
Mexico	0.8	0.2	0.3
Canada	0.8	1.5	2.4
Europe	10.7	14.2	19.1
U.S. other	0.6	0.3	1.3
Unknown	29.3	35.7	23.0
Other	0.9	1.4	1.5

rush accounts for the European influx, while the Midwestern differential may be partly explained by the relative proximity of the region. Also, the number of native-born Californian parents is much higher in the four larger cities (17.5 percent compared with 8.5 percent), because these cities had the colonial Mexican connection before 1850. This was especially true with San Jose, which had 33 percent native-born California parents, while the small towns had an average of just 8.5 percent.

By 1870, the California-born percentages from the large and small cities were more alike. The Midwest was still the second-leading origin for parents after Europe in the small cities. It was nearly the same, percentage-wise, for the large cities, while children born in the small cities were slightly more likely to have California native parents, which was a change. By 1890, the mining towns had become more provincial, with a greater proportion of parents born in California than the large cities (31.6 percent to 26.4 percent), while the large cities had a higher relative share of European parentage. However, the numbers were actually pretty similar, showing the convergence of migration patterns to peripheral and core areas by that time (table 7.5).

INTERTWINED MIGRATIONS

Genealogical geography can rewrite the map of historical migration patterns and relationships around the world. The conceptual generational migration hourglass of communities illustrates underlying diaspora and gathering geographies, which show the relative areal pull of diverse cities and towns during different historical periods. This chapter shows how generational migration reveals different migration processes than basic life-course migration studies, as the latter approach measures only one segment of potential individual movement—birthplace to place of a child's birth. Generational migration thus encompasses a more expansive time scale that is connected by blood over many decades. This allows us to highlight transmission of people and culture over space through intertwined streams of familial relations. The value of this approach then, is its ability to analyze long-term migration patterns around the world. Further, it enables us to describe their character through varied statistical measures and to compare, explain, and reveal past diasporas and gatherings.

The case study provides insight into the numerous historical migration processes encompassed within the larger scale of this chapter. For example, the statistics highlight cities' relative pull from their generational migration hinterland (the source area for the migrants), illustrating that booming mining towns and commercial cities can both attract distant migrants. Also, larger standard distances mean a more geographically diverse incoming population, while a narrow standard distance could indicate chain, religious (Rohrer 2010), or community migration on a larger scale (Durham 1997). Additionally, the ability of the CSI and RCM indices to pinpoint when settlements exhibited long-term stable characteristics is valuable, especially in its comparative ability across geographies and time.

In summary, a significant facet of this chapter's research is its ability to sort out how family relatedness connects to the migrations of larger clusters of people. None of us is an independent and random participant in the world around us. Our migration tendencies and cultural identities are influenced by parents, siblings, and extended family in a multitude of ways (Kok 2010). When we move from one place to another, our culture comes with us and refines and redefines the place where we go. Multiplied numerous times for thousands of people, it is apparent how past cultural flows are inextricably connected within a strong, but often underappreciated, familial structure. To this end, by using a genealogical database that contains hundreds of millions of names, we have expanded the scope of migration analyses to consider the migration behavior of multiple generations.

The complex settlement of northern California received useful explanation in this chapter. People *gathered* to the mining districts and coastal commercial centers from great distances (top half of the migration hourglass; fig. 7.1). Over fifty years, *community stability* increased from very low at the beginning of the gold rush to one that was markedly stable, especially in the mining cities. This highlights the development stage progression from wilderness to firmly established urban hierarchies in the migration and settlement model at this scale. The *genealogical mixing* of the communities showed their cosmopolitan makeup, which decreased in the mining towns over time in comparison with the larger commercial centers.

Exploration of the makeup and movement of those who settled the cities and towns of northern California helps set the stage for more community scale analysis in the next three chapters. Migration is geographically and genetically interconnected, and our sojourn in genealogical history here has helped explain the march of people and culture across the landscapes of California.

8

Hyper-Rate Settlement Evolution and Devolution in the Sierra Nevada

From the start of the gold rush through the ensuing decades of the 1800s, communities up and down the Sierra Nevada transitioned from volatility to stability. As we have seen, the migration attractiveness of the foothill towns rivaled that of the bigger cities to the west. People came to the foothills from all over the world. In this chapter, rather than focusing on instability versus stability, we uncover the geographic factors that brought about differential population change in these settlements—growth versus decline.

There is an important question: why did some mining camps survive, prosper, and develop into cities that still exist and even thrive today, while others faded from memory and the map? Or, how did geography and circumstance combine to foster some settlements and efface others? The conceptual model (fig. 1.2) that provides insight here has similarities to that of multigenerational town attractiveness and "stickiness," with emphasis on the community-level scale, rapid initial migration rates, and underlying natural resources of the geographic interface. However, the focus in this chapter is even more on rapid in-migration and out-migration and what geographic characteristics explain differences from town to town (community-scale analysis), and less on family relationships over time.

The diffusion of mining camps and villages in the Sierra serves as a rapid microcosm of settlement evolution that occurred across the rest of the state. This initial growth and spread occurred at an exponential rate, a pattern very different from regular settlement diffusion because so many of the communities were highly ephemeral. In any settlement system, there are so-called winners and losers. The winners survive and prosper, while the losers shrink and wither. With the Sierra mining towns and camps, most ended as losers, even if millions of dollars of gold were extracted from the vicinity. One need only survey the current landscape of the Sierra and compare it to the period from 1848 to the turn of the century for proof. Many of the boomtowns no longer exist or are only shells of their former aspect.

This chapter explores the booms and busts of the towns and mining camps of four counties in the central Sierra: Nevada, Placer, El Dorado, and Amador. Similar stories could be shared for towns in Calaveras, Yuba, Tuolumne, Sierra counties, and a number of others, but time and space limit analysis.

Placerville, Auburn, Nevada City, and Jackson are four cities in the region that somehow succeeded. Ground Hog Glory, Horseshoe Bar, and Fort John did not. What made the difference? Could it be just as simple as what a man shared with me at the El Dorado County Historical Museum in 2007? He said, "Towns that were on the transportation routes or that were supply centers survived."

The hilly, disjointed geography of the foothills has to be part of the explanation. Arriving in California for an extended stay in 2007, I made an important discovery. My family and I had just traversed Donner Pass and the rest of the high Sierra on a particularly cold day in January. As the evening darkened, we encountered a seemingly endless length of ridge tops and valley undulations, finally arriving in Auburn where the topographical hyperactivity relaxes a bit. Although we were heading west on Interstate 80, our final destination was actually to the southeast near the small town of Pollock Pines, up the hill from Placerville. I was pulling a U-Haul trailer with a twelve-passenger van. Looking at the map before making the trip, the plan was to take Highway 49 from I-80 south to Placerville, as it was the shortest route. We turned on 49, anticipating the end of a long day of travel, only to make the rude discovery that the steep grades and sharp curves of the road were not kind on our vehicle. After a few miles, I turned around, and regained the relative ease of I-80 until we arrived in the sprawling suburbs of Sacramento, where the connection to eastbound Highway 50 toward Placerville was much more doable for a caravan of my size. Later, I had opportunity to drive that curvy road from Placerville to Auburn and reconfirmed I had made a good decision that dark night.

Highway 49 runs through the gold discovery city of Coloma, and its route is at the far opposite end of the spectrum in variation and "white-knuckleness" from any north-south route in the flat, open Sacramento Valley. Today, getting to Coloma is still not easy if you have motion sickness. But it was even more difficult over 160 years ago when Caleb Greenwood found his way there, although "the Coloma Road evolved from a narrow pack trail into a well-beaten, crowded highway soon after Marshall's discovery" (Rensch et al 1932, 64). Most mining camps and towns up and down the foothills were at least as difficult to access because of the region's hilly topography.

South of the January 1848 gold strike in Coloma, another less celebrated discovery occurred in the summer of 1848. As one of many subsequent finds in the region, it occurred upon a spot that would become an enduring town of some consequence: Dry Diggings. But where is Dry Diggings today? Two of the early European settlers of the Central Valley that surrounds present-day Sloughhouse on the Cosumnes River were William Daylor and Jared Sheldon. They both married daughters of Thomas Rhoades (Sarah and Catherine Rhoades), a Latter-day Saint who migrated to California in October of 1846 (Davies 1984, 12; Winkler 2000, 11). These men were purportedly involved in the discovery of gold north and east of Sloughhouse during the summer of 1848 on the north fork of Weber (Weaver) Creek, which became known as "the dry diggings," for which Daylor received the greater credit (Gudde 1975; Winkler 2000, 12). Their find marked the beginning of a town that grew rapidly, especially as it was reported that William Daylor and another companion, Perry McCoon, had extracted $17,000 worth of gold from the diggings in a week (Winkler 2000, 13). Later in the summer the diggings were home to "about three hundred—old pioneers, native Californians, deserters from the Army, Navy and Colonel Stevenson's volunteers" plus Indians, according to James Carson, an early chronicler of the gold phenomenon (Winkler 2000, 19). Thus, a new town was born almost overnight (a.k.a. Placerville). When our friend Clayton Winkler arrived on the scene in 1849, it was a busy place.

Rapid justice mirrored the amazing speed that characterized the growth of Dry Diggings or Old Dry Diggings. A nickname attached itself to this settlement in a macabre way. Hangtown got its toponym from the mob jury that condemned to death three suspected criminals in January 1849. After five men had been tried for robbery of a Mexican gambler named Lopez and received thirty-nine lashes each, three of them were then accused of attempted robbery and murder from the previous fall. The jury found them guilty, and they were soon hung in the middle of town, even though they were being punished for an act that they only tried to commit. John Breen related,

> I saw the men hanged to a tree which tragedy caused the place for several years to be called Hang-town. The lynching even if the men deserved death was barbarous, as it was first decided to flog them which was done, the next day some new evidence came out, and a great number of people collected and hanged them.
>
> I will never forget the day, the ground was covered with snow the skies with clouds all nature seemed to mourn over the cruel act being done. But it may have been for the best. There were no courts, and the power to punish crime had to be with someone. (Winkler 2000, 50 [corrected for spelling])

FIGURE 8.1. State Historical marker at Placerville. Courtesy of the author.

The name Hangtown stuck, although it was never the official one. The spot had other place names too: Gallows Hollow, Stringtown, Middletown, Mormontown, Oregontown, Old Dry Diggings Village, and Dry Diggings City, all of which faded (Winkler 2000, 50). Placerville first found its way onto a map in 1851 (Gudde 1975, 270), and when the town was incorporated, it chose that name over Old Dry Diggings or Hangtown (fig. 8.1). In 1857, Placerville gained the title of county seat from Coloma, which was simply a result of the declining importance of the once-famous gold discovery site. Coloma fell rapidly from 10,000 people at its height to 2,246 in 1860 and only 925 in the 1870 census (Nadeau 1965). Thus, Placerville was the location for rapid settlement, lightning justice, and quickly vacillating name designations as it emerged. This instability paralleled the happenings in the county periphery.

During its first years, Placerville itself was almost ephemeral. Edward Buffum, who lived for a time at Weaver Creek in 1849, wrote later that year that "old 'Dry Diggings' were entirely deserted last spring, having been used as a mere wintering place, and abandoned when the weather admitted of traveling. As immigrants rushed in, however, people again began to settle at the old working-places, and the 'Dry Diggings' were soon again filled up" (Winkler 2000, 52). With this rapid population onslaught in the 1850s, Placerville's population was nearly that of San Francisco or Sacramento.

> Not only was Placerville a mining center, it was also a trade, transportation, and communication center. Because it was situated on the California Overland Trail, Placerville served as an important station for the Central Overland Mail and Stage line, the overland telegraph, and the Pony Express. When the Comstock boomed [after 1860] in Nevada, Placerville became the main stop on the way to the new strikes. (D. Miller 1978, 135)
>
> To this destination specially designed "Washoe wagons," built in Stockton and Placerville, delivered every pound of food, every barrel of whiskey, every machine and pump, all the thousands of flasks of quicksilver...came over two toll roads, one from Placerville, the other from Nevada City. Both those towns skimmed big profits as transfer points from San Francisco and Sacramento, where wholesalers in their turn had taken a fat share. (Holliday 1999, 225)

Areas around Placerville developed extensive underground hard-rock quartz mining operations, but these were not as prolific or long lasting in production as the Nevada County mines. So Placerville suffered when its limited placer and underground deposits gave out, then when the Transcontinental Railroad was completed in 1869, and later as the Comstock Lode in Nevada lost some of its silvery allure (D. Miller 1978). On the other hand, Auburn and Colfax benefited from the completion of the railroad, as they were on its route. Population in Placerville Township declined from 4,700 in 1860 to 2,624 in 1870, and it hovered under 3,000 until after 1920. Some eighty years later, in 2010, Placerville had only grown to 10,389 residents.

Early Jackson, Amador County, had a different name, too. It was first known as Bottileas in 1848 or "place of bottles, so called from the number of bottles lying about the spring" and was a "stopping place for travelers between Drytown and Mokelumne River" (Sargent 1927, 53). Its name was later changed to honor Colonel Aldan Apollo Jackson, who mined there in 1848 or 1849 (Gudde 1975, 172). Mining in the gulches around the town itself wasn't particularly profitable, but underground mining made the difference. The nearby Argonaut (discovered in 1850) and Kennedy (beginning in 1856) underground mines produced millions of dollars worth of gold in their lifetimes and helped to underpin the economy of the city. Dangerous bull and bear fights on Sundays were exciting diversions for the locals in the early days (Sargent 1927, 53). Leaders in the Calaveras county seat, neighboring Mokelumne Hill, didn't approve of such behavior but failed to prevail. Jackson became the permanent county seat of newly formed Amador County in 1854, while "Mok Hill" lost the Calaveras County seat title to San Andreas in 1866.

Being county seat gave Jackson an advantage over neighboring towns such as unfortunate Mokelumne Hill (Gudde 1975). The story of Jackson's and Mokelumne Hill's competition for the county seat title in 1851 illustrates the perceived importance of the designation.

> When the Gold Rush began, Mok Hill got a head start on its neighbor. Discovered in August 1848, it was a thriving, canvas-tented camp within a year.... By the end of 1849 [Jackson's] population was only sixty persons. But as rich diggings were uncovered along Jackson Creek, the camp flourished in its own right.
>
> Then the trouble began. Calaveras County (which then included Jackson) was formed in 1850, and the county seat was won by the nearby camp of Double Springs. Since it had only one building, Jackson and Mok Hill considered this a deliberate insult. So in July 1851, an election was held to relocate the county seat. Mokelumne, still having the larger population, naturally won—or at least that was the result reported by the county clerk, Colonel Collyer.
>
> But the Jackson men were not through. Three of them rode down to Double Springs and invited everybody in the Courthouse (which also served as the saloon) to drink to the new county seat, Mok Hill. This suggestion could hardly be ignored, even by Col. Collyer. And while everybody was toasting, one Jackson man grabbed the official records from a nearby table, threw them in a wagon and dashed off to the new seat of Calaveras County. By the time the county clerk arrived, one Judge Smith had recounted the ballots and declared Jackson the winner. (Nadeau 1965, 78–79)

Another version of the story states,

> [The] first count made Mokelumne Hill the county seat. A later count by Judge Smith gave the election to Jackson. Be that as it may, Charles Boynton and Theodore Mudge appeared early one morning at Double Springs, deluded the custodian, Col. Collyer, threw the county archives into a buggy and took them to Jackson where they set up a court house in a "shack shanty" at the foot of Court Street. (Sargent 1927, 7)

Jackson was only county seat until 1852, when Mokelumne Hill took the official title (from 1852–1866), but fortunate Jackson became seat of the newly formed Amador County in 1854. So Double Springs was the real loser, as the town quickly declined after this political blow, coupled with the end of the easy gold pickings. Its post office closed in March 1860 (Gudde 1975; Salley 1977), while 150 years later in 2010 Jackson boasted 4,651 people and Mokelumne Hill had 646. But at least Mok Hill has more residents than Double Springs!

James Marshall set off from Coloma in 1848 searching for more gold and found a little at Deer Creek. Unfortunately for him, he moved on, which was the story of his life through the end of his days (D. Miller 1978). "Marshall, hounded by miners who presumed he knew where riches could be dug, resorted to selling autographed cards for a paltry income. He died penniless and embittered" in the village of Kelsey near Coloma in August 1885 (Gay 1967, 414; Holliday 1999, 74). The next year, after Marshall left, a man named Hunt had more luck there, and soon a camp called Deer Creek Diggings or Deer Creek Dry Diggings was established. Miners rushed in. Dr. A. B. Caldwell started up a store in October 1849, and the place was then referred to as Caldwell's Upper Store. This name did not last, however, because in March 1850 the growing residential population changed the city name to Nevada (now Nevada City; Gudde 1975, 237). By 1850, the settlement had swollen to 8,000 to 10,000 people and was so successful that it survived being nearly burned to the ground at least five times in its first ten years (D. Miller 1978).

Nevada City and Grass Valley are within miles of each other.

> Both began as placer camps…but they soon diverged economically: Nevada City's growth was founded on hydraulic and quartz mining, supplemented by commerce and county government, while Grass Valley thrived on its hard-rock quartz operations. They survived their more notorious contemporaries because their gold deposits could support the development, by the late 1890s, of large-scale industrial mining economies. (Mann 1982, 1)

Nevada City can claim that it was the biggest mining camp in the state, and at its height only Sacramento and San Francisco had more residents. The city was named after the mountains some eleven years before Nevada Territory was organized, and the locals were none too happy when the new territory copied their name. Now most people associate "Nevada" with the big state next door, not the original named place (Nadeau 1965, 134).

> By 1850 Nevada City consisted of some 250 buildings ("frame houses, dingy old canvas booths and log cabins") extending along the bottom of a ravine. She was hardly a metropolis, but she did have one cultural jewel—a theater.… Isolated as she was, Nevada City early became a great staging center for the Yuba River camps, the Sierra Summit, and points northeast. By 1851 Nevada City stages ran out of Sacramento

FIGURE 8.2. Statue of Claude Chana, California miner and pioneer at Old Town Auburn. Courtesy of the author.

several times a day—carrying express at the rate of $2.50 per letter. (Nadeau 1965, 136)

Nevada City's proximity to its rival, Grass Valley, meant that commerce between the two towns helped keep the connecting road busy.

Auburn, Placer County, followed a script familiar to the other survivors. It was born in the spring of 1848, when a man named Claude Chana and some Native Americans discovered gold northwest of Coloma near the confluence of the North and Middle forks of the American River. During the next year, its name changed from Rich Dry Diggings to North Fork Dry Diggings, and Wood's Dry Diggings. Finally, in 1849, migrants who had arrived from Auburn, New York, named it Auburn. Miners and their pans flooded the place, with about 1,500 working the area in 1850, and it became the county seat of newly formed Placer County the next year (fig. 8.2). But it was only a matter of a few years before the placers began giving out. By 1860, there were only 811 people living in the town. In 1870, there were still about 800 residents, but the town had received the lifesaving arrival of the railroad in 1865 and the full transcontinental connection four years later. By 1890, the population had finally returned to the 1850 level, as county government, transportation, and agriculture took up the slack left by the virtual end of mining (Dilsaver 1986; Gudde 1975; D. Miller 1978).

Auburn has a distinctive historical geography, being a "three story town." At the lowest story is the location of Auburn's first gold discovery. "The new locality produced as much as $1,500 a day for a lucky few. Of this original site little is left today save some half-hidden foundations" (Nadeau 1965, 56–57). The second story is located nearby at the place where the old town was laid out with a Spanish plaza design. "On its narrow and winding streets stand old-time buildings of false fronts and wooden canopies." The top story is higher still and is the center of modern Auburn (Nadeau 1965, 56–57).

These three stories actually represent the evolution of many gold-rush towns from nearly vanished early mining remains, historical buildings of the 1850s and 1860s (the few still standing), to the much larger modern district. In the case of the smallest mining camps, nothing remains but the first story. Some of the more roaring mining towns that have descended to ghost town status might have both the first and second stories extant. It is only places such as Auburn, Placerville, Jackson, and Nevada City that actually have a thriving third story, even if it is not physically at a higher elevation like Auburn's.

I analyzed population growth and decline and employment structure in a sample from towns in Nevada, Placer, El Dorado, and Amador counties for 1860, 1880, and 1900, by comparing towns that are now very small (Volcano, Gold Run, Forest Hill, and Georgetown) with Nevada City, Placerville, Auburn, and Jackson (tables 8.1a–d). It is not easy to differentiate the towns that survived and thrived from those that did not using employment structure, but the bigger towns relied less on the primary sector by 1900. The small towns couldn't seem to move away from mining and other resource extraction activities.

FOOTHILL SETTLEMENTS

Beginning at the 1848 gold strike in Coloma, El Dorado County, Lary Dilsaver considered the diffusion and dissipation of mining camps along the north to south line of the Sierra foothills and mountains (Dilsaver 1982, 1983). He illustrated the amazing speed of this process. Dilsaver's research underscores the locational proximity of the approximately 191 towns in 1855 to certain elevations in the foothills and rivers, especially the Middle Fork of the American River between Placer and El Dorado Counties. As the number of settlements shrank to just eighty-two by 1880, the decline of placer mining was very evident. Additionally, the remotely situated river settlements of 1855 were nearly eliminated by 1880. Erwin Gudde's (1975) summary work lists these places and when they appeared on maps and in other documents (table 8.2).

TABLES 8.1a–d. Sector employment, derived from the 1860 and 1900 U.S. censuses for the towns of Nevada, Placerville, Auburn, Jackson, Volcano, Gold Run, Forest Hill, and Georgetown.

TABLE 8.1a

1860	NEVADA		PLACERVILLE		AUBURN		JACKSON	
SECTOR	#	%	#	%	#	%	#	%
Primary	110	63	51	34	40	22	126	67
Manufacturing	13	7	29	19	25	14	9	5
Services	33	19	59	39	100	54	33	17
Transportation	2	1	0	0	5	3	0	0
NA*	4	2	0	0	1	1	1	1
Laborer	12	7	11	7	14	8	20	11
Miner	109	63	43	29	32	17	114	60
Farmer	0	0	6	4	2	1	7	4

*NA—Not active in economy #—Number of people %—Percentage of total

TABLE 8.1b

1860	VOLCANO		GOLD RUN 1860–70		FOREST HILL		GEORGETOWN	
SECTOR	#	%	#	%	#	%	#	%
Primary	135	69	94	72	91	53	134	65
Manufacturing	19	10	8	6	17	10	16	8
Services	26	13	22	17	51	30	53	26
Transportation	1	1	3	2	9	5	1	0
NA*	2	1	0	0	1	1	0	0
Laborer	13	7	3	2	3	2	1	0
Miner	126	64	90	69	91	53	131	64
Farmer	7	4	4	3	0	0	1	0

Gold Run is from combined 1860 and 1870 censuses
*NA—Not active in economy #—Number of people %—Percentage of total

TABLE 8.1c

1900	NEVADA		PLACERVILLE		AUBURN		JACKSON	
SECTOR	#	%	#	%	#	%	#	%
Primary	53	48	39	38	23	22	66	56
Manufacturing	11	10	9	9	18	17	11	9
Services	34	31	36	35	54	51	31	26
Transportation	7	6	3	3	2	2	3	3
NA*	0	0	9	9	0	0	0	0
Laborer	5	5	6	6	8	8	6	5
Miner	51	46	21	21	18	17	50	43
Farmer	0	0	16	16	4	4	13	11

*NA—Not active in economy #—Number of people %—Percentage of total

TABLE 8.1d

1900	VOLCANO		GOLD RUN		FOREST HILL		GEORGETOWN	
SECTOR	#	%	#	%	#	%	#	%
Primary	68	59	23	59	85	70	69	61
Manufacturing	5	4	4	10	7	6	10	9
Services	23	20	7	18	22	18	7	6
Transportation	4	3	4	10	5	4	3	3
NA*	0	0	0	0	0	0	0	0
Laborer	16	14	1	3	3	2	25	22
Miner	46	40	19	49	69	57	44	39
Farmer	22	19	4	10	15	12	25	22

*NA—Not active in economy #—Number of people %—Percentage of total

TABLE 8.2. Mining camps and towns by years listed by Gudde (1975) in Amador, El Dorado, Nevada, and Placer counties.

COUNTY	BEFORE 1848	1848	1849	1850	1851	1852	1853–59	1860–69	1870+	NO DATE
Amador	0	18	24	31	14	15	52	28	13	19
El Dorado	1	33	49	67	29	20	122	48	71	0
Nevada	2	2	32	61	22	17	112	93	58	51
Placer	2	9	31	46	29	44	132	67	44	0

The rapid surge of population on the western foothill slopes of the Sierra resembled the powerful and overwhelming rush of a tidal wave battering the seashore (D. Miller 1978, 2). People flooded the land, but after a brief period most drained away, leaving once-thriving towns and mining camps only shells. Some, however, were fortunate enough to be situated in locations that enabled them to stay due to more plentiful gold sources, successful economic adjustment, or a central trade location. These "tide pools" of people contrasted with the decline in the less fortunate locales surrounding them.

Population waves swept the landscape in various places and years, but the result was similar each time: some towns persisted and even prospered, while other places collapsed into oblivion. The effects of these erosive human rushes left imprints on the state's natural environment, which included the far northern Trinity and Siskiyou region that expanded in fits and starts concurrent with the Sierra mining rush, and the eastern Sierra region of Bodie and Mammoth City of the 1860s and 1870s. The rise and fall of these communities illustrates how diverse local geographies contributed to success or failure and is still reflected in the landscape of California.

To complement an economic structure analysis, I constructed a network grid of the post offices for 1851, 1856, 1875, and 1889 after Hong (2009). Post offices are one indicator of town viability. "The establishment and discontinuance of post offices reflect the rise, ebb, and distribution of the people and material resources of our country" (Salley 1977, preface). If a town gets large enough to warrant local mail service, then it follows that the length of time the post office stayed open can help delineate the town's lifespan.

The towns that survived and that have sizeable populations today were the ones that, very early on, fulfilled a variety of economic functions and thus were less dependent on mining. They also showed centrality by the importance level of their post offices. In 1851, just after statehood, there were some twenty-one post offices in the state, with Culloma (Coloma), Placerville, and Nevada City being in our study area, highlighting the relative importance of this region imparted by gold mining. These three post offices multiplied to

TABLE 8.3. Post offices, 1856: Amador, El Dorado, Nevada, and Placer counties with postmaster salary and office revenue (source: United States 1875).

COUNTY	POST OFFICE	ANNUAL SALARY $	ANNUAL REVENUE $
Amador	Drytown	827.58	187.88
	Jackson	648	889
	Volcano	749.86	485.08
El Dorado	Cedarville	74.49	46.6
	Cold Spring	279.19	166.74
	Culloma or Coloma	1,583.19	1,474.26
	Diamond Springs	839.34	504.95
	Fiddletown	242.45	128.51
	Garden Valley	35	19
	Georgetown	1,240.44	1,014.34
	Green Wood	454.63	303.54
	Indian Diggin(g)s	115.8	54.92
	Newtown	57	34
	Pilot Hill	93.11	46.87
	Placerville	2,500	2200
	Salmon Falls	185.6	101.51
	Spanish Flat	277.51	166.49
Nevada	Clinton	110	50
	Grass Valley	1,462.8	1,245.48
	Nevada City	2,000	2,248.99
	Rough and Ready	733.48	290
Placer	Auburn	846	627
	Illinoistown	165.61	58.17
	Iowa City	720	487
	Michigan Bluff	843.52	190.88
	Mountain Springs	260.67	84.24
	Ophirville	422.05	235.41
	Oro City	108	27
	Rattlesnake Bar	197	42
	Secret Ravine	155	50
	Yankee Jim's	784.99	491.06

forty-six during the next five years in the four counties as mining boomed. Revenue for 1856 by post office indicates relative importance of the different towns (table 8.3; fig. 8.3). Nevada City and Placerville were perched on the top level of the hierarchy, each with incomes around $2,200. Coloma, Grass Valley, Georgetown, and Jackson formed the second tier in descending order. Below this were the more minor communities. Since then, most of the third tier have shrunk to ghost towns or quiet remnants with few inhabitants.

Yankee Jim's was a representative boom and bust town. Gold was discovered in the area in 1850 by a reputed rowdy, Yankee Jim (also known as Jim Hill or an Australian called Robinson). He was a horse rustler, but it turned out that the corral where he kept his stolen animals also harbored gold. According to the story, a mining camp sprang up in the corral area, and by 1857 it was "one of the most important [towns] in Placer County." It was also an early site for the innovation of hydraulic mining (*California Historical Landmarks* 1996, 159; D. Miller 1978, 180). In 1856, it was doing quite well, and post office annual revenue was $491. But the gold played out and the town slowly faded until the post office itself was closed in 1940. In 1965, one writer said, "Once a large mining town, Yankee Jim's is marked today by a few old buildings in a delightfully shaded dell" (Nadeau 1965, 62). At this writing, Yankee Jim's is a country road, and its namesake ghost town is so "ghostly" that it has little to nothing to show for itself.

By 1859, there were some sixty post offices in these four counties of the Sierra. Less than ten years before, a man could only get his mail in three towns. Growth was substantial but slowed after 1859, as population stabilized or even declined (fig. 8.4). Even before then, some places had already seen post offices come and go, including Butte City, Amador County, which opened its post office in 1857 only to close it in 1858, even though mining was still underway in the area until at least 1866 (Gudde 1975). Butte City is now a ghost town along Highway 49 (fig. 8.5). And the world probably didn't notice how many other post offices, such as Lone Star, Jay Hawk, Oro City, Bottle Hill, and Cedarville, dropped off the map in the decades after 1855.

In 1875, there were just sixty-seven post offices in the region. The town of Coloma continued to be important for a while after 1856, but by 1875 it had faded and by 2010 had only 529 residents. In 1875, Grass Valley was classified as a First Class post office; Nevada City was Second Class; and Placerville and Truckee were Third Class. The salaries for their postmasters were $3,000, $2,100, $1,300, and $1,300 respectively. Amador County had only Fourth Class post offices (United States 1875). Coloma was physically out of the way of the main east-west transport routes, so its shrinking importance after the gold placers gave out is logical. Nor was Georgetown able to keep up. It was off the east-west transport routes, and its economy sputtered as the gold played out (2010 population: 2,367). Grass Valley and Nevada City were the most important cities at the time as they had outlasted the placer mining decline by moving to hydraulic and then deep quartz gold mining (Mann 1982; D. Miller 1978). Over the years, Grass Valley has eclipsed Nevada City, even though Nevada City is the county seat (12,860 versus 3,068 people, respectively, according to the 2010 census; fig. 8.6).

Of the four study counties, El Dorado County had the most precipitous loss of residents: it was nearly halved between 1860 and 1870 (20,562 people

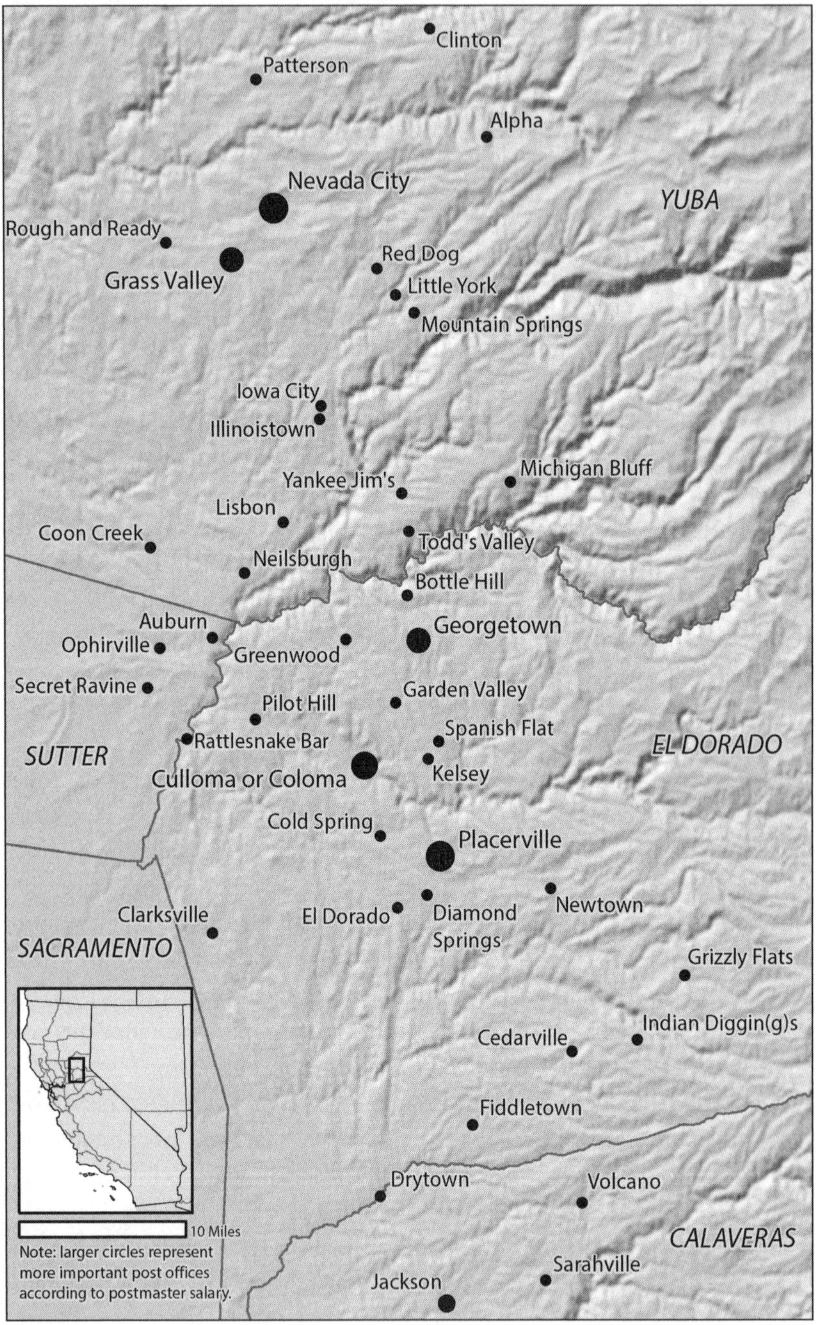

FIGURE 8.3. Post offices, 1856. Data sources: United States 1875; Salley 1977. Marcus Allen Romboy and Jeffrey Stark, ThinkSpatial, BYU Geography.

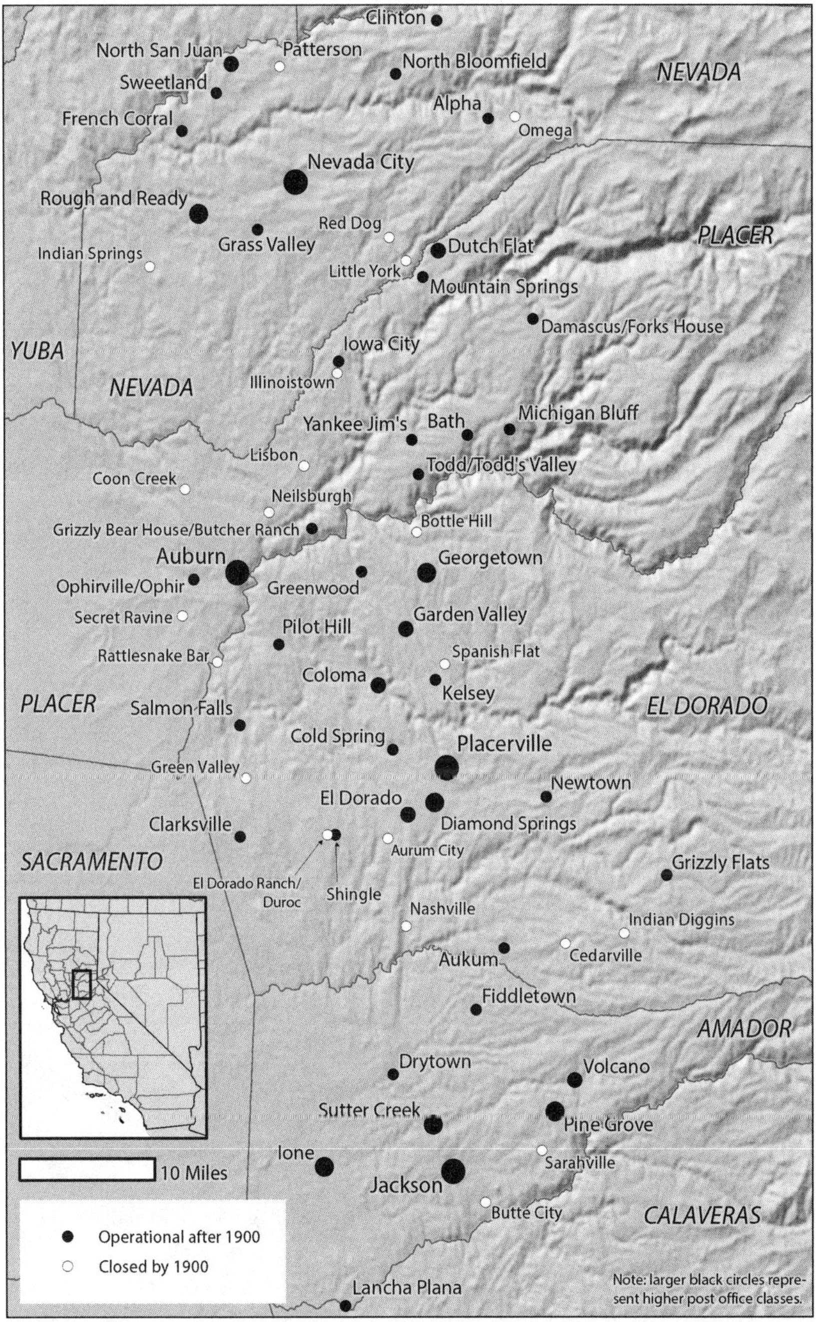

FIGURE 8.4. Post offices opened before 1860 and operational status, 1900. Data source: Salley 1977. Marcus Allen Romboy and Jeffrey Stark, ThinkSpatial, BYU Geography.

FIGURE 8.5. Butte Store, last vestige of Butte City. Courtesy of the author.

FIGURE 8.6. Relics at Empire Mine State Historic Park, Grass Valley. Courtesy of the author.

TABLE 8.4. Amador, El Dorado, Nevada, and Placer county populations 1860–1950 (U.S. Census).

	AMADOR	EL DORADO	NEVADA	PLACER	TOTAL
1860	10,930	20,562	16,446	13,270	61,208
1870	9,582	10,309	19,134	11,357	50,382
1880	11,384	10,683	20,823	14,232	57,122
1890	10,320	9,232	17,369	15,101	52,022
1900	11,116	8,986	17,789	15,786	53,677
1910	9,086	7,492	14,955	18,237	49,770
1920	7,793	6,426	10,850	18,584	43,653
1930	8,494	8,325	10,596	24,468	51,883
1940	8,973	13,229	19,283	28,108	69,593
1950	9,151	16,207	19,888	41,649	86,895

down to 10,309). It continued its downward slide until the low point of 1920 at 6,426 people. Depopulation in the mining camps and in the cities, caused by the downfall of mining and shifts in transportation, changed the whole face of the county from booming center to sleepy outpost. Placerville's county seat status and location along some major roads allowed it to survive as the largest town in the county, but it too was a shell of its former self as the railroad bypassed it and gold sources were depleted. A post office city that was not a gold mining town was Truckee, Nevada County, which grew to a population of 1,147 in 1880 owing to its location along the Transcontinental Railroad and its ice harvesting industry (Truckee History 2013).

In terms of population numbers, the 103 post offices in these counties in 1889 seem unjustified. True, the four counties had added several thousand people since 1870 to be in the range of 52,000 people, but it could also be that the U.S. Postal Service was working to extend its reach to smaller communities in the late 1800s (table 8.4). The number of people per post office at that time was approximately 500 (compared with approximately 880 people per post office in 1859). Regardless, after 1889, the post offices in this region steadily closed. Some thirty-five of the 103 pre-1890 post offices went out of business from 1890 to 1930, and an additional fifty-seven were opened and closed during the four decades (fig. 8.4). By 1976, there were approximately ninety-four post offices in the four counties, nine less than in 1889 (Salley 1977). After 1890, new post offices in these counties were opened and old ones closed in tandem with the changing fortunes of the region, whether it came from mining, recreation, agriculture, or transport. Towns that retained their post offices throughout the period usually had locational advantages along primary roads, had diversified their economies, or had found gold underground.

What was more fleeting than the expansion of the post offices was the fate of the rest of the mining camps and even smaller towns in the region. *California Gold Camps*, by Erwin G. Gudde (1975), offers an impressive sense of the tremendous number of settlements and temporary camps that exploded in California's mountains beginning in 1848. The places in his book range from prospecting sites to temporary mining camps to enduring larger towns. In the four-county study area, there were hundreds of places that appeared and mostly went out of existence during the last half of the 1800s. There were about 214 of these sites in Amador County, 440 in El Dorado County, 450 in Nevada County, and 404 in Placer County. More impressive is that a majority of these locations were founded and named by 1855 (at least 278 in Placer, 135 in Amador, 273 in El Dorado and 224 in Nevada). This is the period when the number of post offices in those counties increased from three to forty-six between 1851 and 1856. Most of these mining locales were ephemeral and never remained occupied long enough to merit post offices.

Their names tell much about the hopes and character of the miners themselves. In Nevada County, there was Lola Montez Diggings, named after the famous actress and dancer who came to Grass Valley from Europe bringing her strange Spider Dance and latest husband with her. She was from Ireland originally, but had changed her persona to that of a Spanish dancer. Earlier, she had attracted the likes of Ludwig I of Bavaria and became his mistress. After her popularity started to decline in Europe, she came to America and performed in various large cities, finally trying her luck in San Francisco, Sacramento, and eventually Grass Valley. There, she kept pet grizzly bear cubs and developed a busy social life that was not amenable to keeping a husband. She did train a young woman in town, Lotta Crabtree, to become a dancer and singer. Lotta became successful in her own right, before Lola left town for Australia. Within six years' time Lola came to an ignominious end in New York (Nadeau 1965).

Animals of all types made their way into the local toponyms. Gudde (1975, 171) states, "Of the names of domesticated animals [jackass] was by far the most popular for camps and mines." There was a Jackass Flat in El Dorado County, a Jackass Gulch in Amador, and Nevada County had its own Jackass Gulch. Add to that Muletown in Amador, plus Donkeyville and Mule Springs in Nevada County, and these related creatures rise even higher in geographic importance. The local danger of rattlesnakes was emphasized in Rattlesnake Bars in each of El Dorado, Placer, and Nevada counties. A more imposing creature was immortalized in Grizzly Hill in Nevada County; Grizzly Hill in Amador; Grizzly Flat, Grizzly Bar, and Grizzly Bear House, all in Placer County; and Grizzly Flats in El Dorado. Grizzly Bear House and Grizzly Flats grew to have their own post offices, with Grizzly Bear House lasting from 1858 to 1871. Grizzly Flats continues today as a small community of about 400 in an out-of-the-way corner of El Dorado County. Other common animals are remembered in Eagle City (El Dorado), Coon Hollow and Coon Hill (El Dorado), Woodpecker Ravine (Nevada), Deer Creek (Nevada), and Skunk Gulch (Placer). Finally, Bedbug, Amador County, which later became Ione, and Mosquito Valley, El Dorado County, remind us of the unpleasant little creatures that often afflicted the miners. Unfortunately, but perhaps not surprisingly, only one of these many animal places (Grizzly Flats) still exists with its original name (Gudde 1975).

Optimism reigned when names were put in place. In the four counties, there were at least nineteen camps, mines, or towns with "gold" somewhere at the beginning of the place name. The Mormons named Pleasant Valley (El Dorado) on their way east out of California in 1848, while Camp Beautiful (Nevada) was designated by an A. J. McCall in 1949, though he didn't bother to record its location. On the other hand, there is only one Fools Hill, in Nevada County, while there is Shenanigan Hill in Placer County, where "gold in the seam diggings here appeared irregularly in pockets, and miners were said to have made nothing or a fortune" (Gudde 1975, 317). One wonders about the morals of the one who named Gomorrah, Nevada County, or the drinking habits of the founder of Bourbon Hill, Nevada. In the case of Whiskey Flat, Amador, "In 1855 Andy Karney mined here with a partner. One of them fell into a hole while slightly intoxicated and discovered a pocket of rich pay dirt" (Gudde 1975, 369).

Colorful names such as Bean Poker Flat, Gold Run, Ground Hog Glory, Horseshoe Bar, Dead Man's Flat, and Condemned Bar are common among the many other places that covered the ravines and flats and riverbeds. Some places have a slightly more lengthy history, while others were just a momentary blip on someone's record or map. Condemned Bar is now under Folsom Reservoir, but in its heyday in 1849 it had 200 miners working the American River. By 1852, there were some fifty miners taking out some five dollars of gold each day. It was still there in 1855, but not as busy as before.

There were two Horseshoe Bars along the American River: the lower one was first mined by the Mormons in 1848, and the upper one was the early workplace for Andrew Smith Hallidie, "later of San Francisco cable car fame" (Gudde 1975, 160). In 1855, at the age of nineteen he built a 220-foot-long aqueduct and a suspension bridge at the site. Even with these innovations, the Horseshoe Bar settlement didn't last long. Bean Poker Flat (on the American River) and Dead Man's Flat (Amador County), on the other hand, are barely worth mention, as little is known about their gold history, but we do know that three dead men were supposedly found at the site of Dead Man's Flat! And then there is Ground Hog Glory. It vanished sometime after 1860 without much of a trace, even though it was deemed a settlement in earlier years. However, it is likely that large numbers of ground digging rodents still live in the vicinity!

Back on the successful side of things, Gold Run in Placer County had some exorbitantly priced real estate in the heady days of 1850—up to $18,000 per claim. Miners developed a technically impressive hydraulic operation there, washing about 80 million cubic yards of gravel using the process, which produced 6 million of gold that exited the city's express office between 1865 and 1878. It had a census population of 377 in 1880. Some mining continued until 1915, but it was minimal after hydraulic mining was banned (*California Historical Landmarks* 1996; Gudde 1975, 136). Gold Run is now a quiet community along I-80, a convenient location that has helped it survive on the map.

More recently, both Placer and El Dorado counties have experienced significant growth over the past decades because of proximity to the Sacramento metropolitan area. Nevada and Amador counties have grown too, but at slower rates. The former forty-niner Shangri-la has been transformed into a retirement and recreation destination. Skiing, hiking, milder temperatures, mountain vistas, and the distinct gold country history attract visitors and new residents alike. Although some mining and logging continue, the region has been reinvented by a new emphasis on enjoying and protecting the environment rather than extraction. The unparalleled transience of the communities of the Sierra during the last half of the 1800s has been replaced by a more sustained growth and settlement durability over the last several decades.

There is truth to the man's simple statement in the historical museum that transportation routes helped channel settlement success to certain towns once the gold rush dissipated. Placerville did not receive the railroad, unlike Auburn, but at least it was on an important route over the Sierra. It is also the county seat and in 2010 had a population of 10,389. Auburn (13,330 people in 2010) was sustained after its gold height by arrival of the railroad, being the Placer County seat, and then the highway (now I-80) over Donner Pass. Both of these cities are also relatively close to Sacramento. On the other hand, the mining towns of Volcano and Fiddletown were not on important trans-Sierra routes. Volcano had shrunk to just 115 inhabitants in 2010, but in its day in the 1850s, it was the seat of Amador County. Fiddletown now has about 235 people. Its claim to fame is tied to a large fiddle on a local building and Bret Harte's "An Episode in Fiddletown" (*California Historical Landmarks* 1996; Gudde 1975, 115). Both are nice, quiet communities, and will most likely remain that way.

9

Historical Geography of Transportation and Development in Lake County

At each spatial level, there is a narrative that places people and environment in an intricate relationship. The race by individuals and families to develop useful livelihoods while establishing communities underscores that relationship. This chapter builds upon the findings of the previous analysis of towns that survived and those that did not in the Sierra Nevada. The environment's effect on settlement development is analyzed at a small regional level, and the interrelationships of physical geographic factors and changing economic conditions both inside and outside the region that help explain settlement growth are highlighted. Evolving transportation methods, as well as barriers to movement caused by physical features, were crucial elements of the settlement story of California itself, and here their importance will be shown at a more local level (Otterstrom 2003b).

Citizens of aptly named Lake County still made their living on the land in the latter half of the 1800s, but instead of primary reliance on gold mining, as with the Sierra Nevada communities, the variety of economic activities and opportunities there was much greater. Because the environment is so crucial to understanding the course of development, the history surrounding this place must not be separated from its underlying geography. In this chapter, I show the centrality of the geographic base component, which is the interface between the functional and spatial perspectives in the migration and settlement model (fig. 1.2). In other words, exploration at the community (county) scale allows us to see how geographic relationships affected the functional growth rate in regard to access, amenities, resources, and capital inputs that drove Lake County's settlement growth during the latter half of the 1800s.

Getting to Lake County, California, has never been easy. Robert Louis Stevenson, in his book *The Silverado Squatters*, summed up this fact with his 1880 description of the effort required to reach the south end of the county.

> This pleasant Napa Valley is, at its north end, blockaded by our mountain [St. Helena]. There, at Calistoga, the railroad ceases, and the traveler who intends faring farther, to the Geysers or to the springs in Lake County, must cross the spurs of the mountain by stage. Thus Mount Saint Helena is not only a summit, but a frontier; and up to the time of writing it has stayed the progress of the iron horse. (Stevenson [1884] 1972, 10–11)

This is the story of the central role that transportation improvements played in the growth and development of Lake County in northern California. These changes, coupled with physical advantages, led to the county's emergence from remote obscurity to a place that is still relatively "out of the way" but has surpassed the population density of its eastern and western neighbors. The relative remoteness of Lake County from the urban influences of San Francisco, Sacramento, and Santa Rosa is due to its distance and distinct physical geography. An appreciation of this particular geography is central to understanding the effect of transportation developments on its settlement history.

SETTLEMENT FACTORS

A general rule of settlement potential is that the closer a place is to a population center, the greater the propensity for settlement. This basic premise can be greatly modified by the presence or lack of a number of important characteristics. For example, physical barriers, such as mountains, swamps, and large rivers, can impede settlement. Scarce or valuable resources, agricultural potential (e.g., good soils, beneficial climate, market access), and certain amenities increase an area's attraction. Transportation improvements can facilitate settlement and work toward equalizing the negative effect of physical barriers and distance. Here, I examine how the combination of these factors, especially in relation to transportation eras, influenced the settlement growth of this fascinating California county.

PHYSICAL GEOGRAPHY

Most of the area of Lake County was originally attached to Napa County to the south. Once settlement began to accelerate in the valleys of the northern section of Napa County, the need for a new county was recognized. In 1861, the state legislature approved a petition to form a new county from part of Napa. It was named "Lake" because of its many lakes, especially Clear Lake, which dominated the central portion of the new county. It was now a completely separate entity, both legally and physically (Coy 1923, 25).

Lake County was given well-defined borders that were directly tied to its physical geography. The county is within the very topographically diverse Coastal Range of northern California, and although its mountains rarely reach extreme heights, successive ridge and valley systems make the region more difficult to traverse than one might suppose. On the western side of the county down to Napa are the Mayacamas Mountains. To the east are the mountains known historically as the Bear Range. These two ranges converge in the northern portion of the county in a rugged series of mountains and valleys that are now mostly national forest and that include the highest peaks of the county (over 7,000 feet). The topography in the southeast is not as high in elevation as other parts of the county, but what it lacks in height is made up in difficulty of traverse because of the rugged topography (fig. 9.1).

In the center of all these barriers is Clear Lake. It is some 19 miles long and 8 miles wide at its widest point. It is also the oldest natural lake in North America and the largest natural lake in California. Clear Lake drains through Cache Creek, which follows a fairly tortuous course to the Sacramento Valley. The valleys that surround the lake (especially Big Valley) contain the greatest amount of arable land in the county. As a result, population has been concentrated between the lake and the mountains. In essence, the central portion of Lake County is a large basin that is attractive due to its beautiful lake and fertile soils, not because of any ease of ingress or egress.

EARLY EXPLORERS

Clear Lake, known by the Indians native to the area as *Kabatin* ("Big Lake"), is not an easy place to find, even though it is located in the center of the county. Indeed, settlement in the area came relatively late compared with neighboring regions. In the early days of exploration and settlement, Clear Lake was mentioned by mountain men and Spanish ranchers. Ewing Young, a very active fur trader in the Southwest, passed it in 1833 on his way to the coast from the American River region. Sometime around 1840, Mexican Captain Salvador Vallejo began a cattle ranch centered on what is now Kelseyville. The operation lasted for a number of years "until the valley was full to overflowing with cattle, which had long since become as wild as deer, and about as hard to handle" (Palmer [1881] 1995, 49). Vallejo took as many cattle as he could with him when he left the area and sold the rest to some new settlers, Mr. Stone, Andrew and

FIGURE 9.1. The rugged hills separating one from Clear Lake and Mt. Konocti (in the background) on Highway 175 between Lakeport and Hopland illustrate the travel challenges early settlers faced in Lake County. Courtesy of the author.

Benjamin Kelsey, and Mr. Shirland (or Shirley). The Kelsey brothers arrived in California from Kentucky as part of the early 1841 Bartleson Party, which was the first overland pioneer company to enter California (Stewart [1962] 1983).

An illustration of how easy it was to get lost during the pre-European settlement phase comes from an account by Edwin Bryant in 1846. He had trouble traveling to the Napa Valley from Sutter's Fort and was led astray by a Native American.

> I knew that we had been traveling out of our course all day, and it was now three o'clock, PM. Rain and mist had succeeded each other, and the sun was hidden from us by dark and threatening masses of clouds.

> We had no compass with us, and could not determine the course to Nappa [sic] valley or Sonoma. Believing that the Indian would have some knowledge of the latter place, we made him comprehend that we wished to go there, and inquired the route. He pointed in a direction which he signified would take us to Sonoma. We pointed in another course, which it seemed to us was the right one. But he persisted in asserting that he was right. After some further talk, for the shirt on my back he promised to guide us, and placing a ragged skin on one of our horses, he mounted the animal and led the way over the next range of hills. The rain soon poured down so hard upon the poor fellow's bare skin, that he begged permission to return to which we would not consent; but out of compassion to him I took off my overcoat, with which he covered his swarthy hide, and seemed highly delighted with the shelter from the pitiless storm it afforded him, or with the supposition that I intended to present it to him.
>
> Crossing several elevated and rocky hills, just before sunset, we had a view of a large timbered valley and a sheet of water, the extent of which we could not compass with the eye, on account of the thickness of the atmosphere. When we came in sight of the water, the Indian uttered various exclamations of pleasure; and although I had felt but little faith in him as a pilot from the first, I began now to think that we were approaching the Bay of San Francisco. (Bryant 1848 quoted in Kelly and Morgan 1965, 222–23)

Yet Bryant found that he was not viewing San Francisco Bay but Clear Lake. He and his companions next encountered Caleb Greenwood, whom we have already discussed in previous chapters, and a number of other mountain men who were able to give him valid directions. This story illustrates the fact that Clear Lake was both relatively remote and undeveloped but not untouched by outside influences by the mid-1840s.

The pleasant existence of the Native Americans who lived around Clear Lake (Pomo and other Indians) was soon to end. In 1847, Andrew Kelsey and Charley Stone were among those who purchased Salvador Vallejo's remaining cattle and grazing rights. Stone and Kelsey were very unkind to the Native Americans to the point of enslaving them for hard labor on the ranch for nearly two years. Then things got worse. The Pomo Indians were recruited by Ben Kelsey (brother of Andrew) to participate in a failed gold mining expedition in the Sierra. They were not paid, and many fell ill with malaria. Death by disease, hunger, and possible conflicts with the Corusia (Colusa?) Indians, their enemy, on the return trip meant that only a few of the original gold-seeking Indians ever returned to Clear Lake. In the fall of 1849, angry Indians murdered Andrew Kelsey and Charley Stone (Palmer [1881] 1995, 55–56).

The following year, the U.S. Army took vengeance. A Lieutenant Lyons and his soldiers traveled over the mountain from Napa via Pope and Coyote Valleys to punish the Native Americans. They soon realized that the Pomos had retreated to a small island in Clear Lake near the present town of Upper Lake. Lieutenant Lyons sent for help from Benicia (near San Francisco Bay) and received it in the form of two whaleboats and two cannons. "These boats were brought up on wagons, and those wagons were the first ever seen in Lake County" (Palmer [1881] 1995, 57). The soldiers then proceeded to kill the Indians (Nixon 1966, 80).

Although the Army was responsible for the wrongful killing of many people, they effectively opened the region to further settlement by building a "road" from Napa, the successful import of wagons, and a peace treaty subsequent to the bloodshed on the island. The road "was the only thoroughfare for wagons until 1865" (Mauldin n.d., 7105). The area was now poised to grow more significantly as the southern portion of Napa County had already begun to do.

THE ATTRACTION OF CLEAR LAKE AND ENVIRONS

Despite the difficulty of travel to Clear Lake, word got out that it was a region of opportunity. With only one road and a number of trails, settlement still proceeded. By the 1860 census, the population of the Napa County area that would soon become Lake County had grown from a few families in 1850 to some 964 residents. This number increased greatly as better roads were built, and in 1861 Lake County was officially created.

After 1850, two of the earliest products to be exported from the county were cheese and hogs. Cheese was valued by weight and could be transported to distant cities without rotting. For a time, Lake County cheese was reputed to be the best in the state. In those early years, at least seven cheese factories were established. Hogs could be driven to market, and these animals were raised to sell in areas as far away as the gold mines in the Sierra. Trips to these markets were not only important for the early settlers' incomes but also provided an opportunity to purchase goods that were not readily available in their frontier communities.

The agricultural statistics from the county assessor's records indicate that some 26,000 hogs and 23,056 pounds of cheese were produced in the county in 1868, but success was short lived. By 1880, other, potentially more lucrative crops began to emerge. Consequently, hog and cheese production dropped to just 4,539 and 7,200 pounds, respectively (Palmer [1881] 1995, 87). Agricultural production remained in flux over the years, as Lake County farmers experimented and searched for the crops best suited to local environment and transportation constraints.

The crops that eventually became the most important were pears and to a lesser extent walnuts and beans. A fine climate combined with other factors gave these products an advantage. Plantings of beans by a Mr. Henry Wambold in 1899 in fertile wetland soil near Blue Lakes was the beginning of a bean producing and canning industry that lasted for many years. Commercial walnut orchards began operating in the early 1900s. They were encouraged by earlier experimental orchards that produced some good nuts, as well as some really poor ones. Grafted varieties were later found to be the best for quality and quantity, and outside opinion encouraged more planting. Walnuts stored well and could therefore be transported out of the county successfully. Only pears outpaced walnuts in production.

Pears became the crop of primary significance by a fortuitous event, for they were just one of many different pioneer fruit crops in the county. In 1885, the California committee that was charged with sending prime examples of produce from the state to the World's Fair in New Orleans took some pears and apples from Lake County. The Bartlett pears impressed the exhibition committee, and one of its members purportedly remarked, "[I]f he were a young man he would buy land in Lake County and plant Bartlett Pears" (R. Lewis and Benson 1949, 71). Soon, commercial pear orchards were being planted there in large numbers.

It did not take long for farmers to realize that it was difficult to preserve pears on the trip from the orchard even to close markets without refrigeration. The problem, of course, was exacerbated by the county's lack of a railway and difficult mountain roads. The solution that helped overcome this obstacle was the development of fruit drying, which allowed transport without decay and at a much-reduced weight. Dried fruit became very popular, as the best grades of pears were dehydrated and sold in various cities. By 1922, the county produced some 1,000 tons of dried pears, and Kelseyville, near Lakeport, boasted the world's largest pear-drying yard.

Pears continued to be a very important crop throughout the twentieth century. However, dried pears lost their edge. After paved highways were built through the county, fresh and canned fruit could be more easily transported out of the area. From 1970 to 1975, income from pear production equaled 57 percent of the county's total agricultural output, with 56.5 percent of that crop

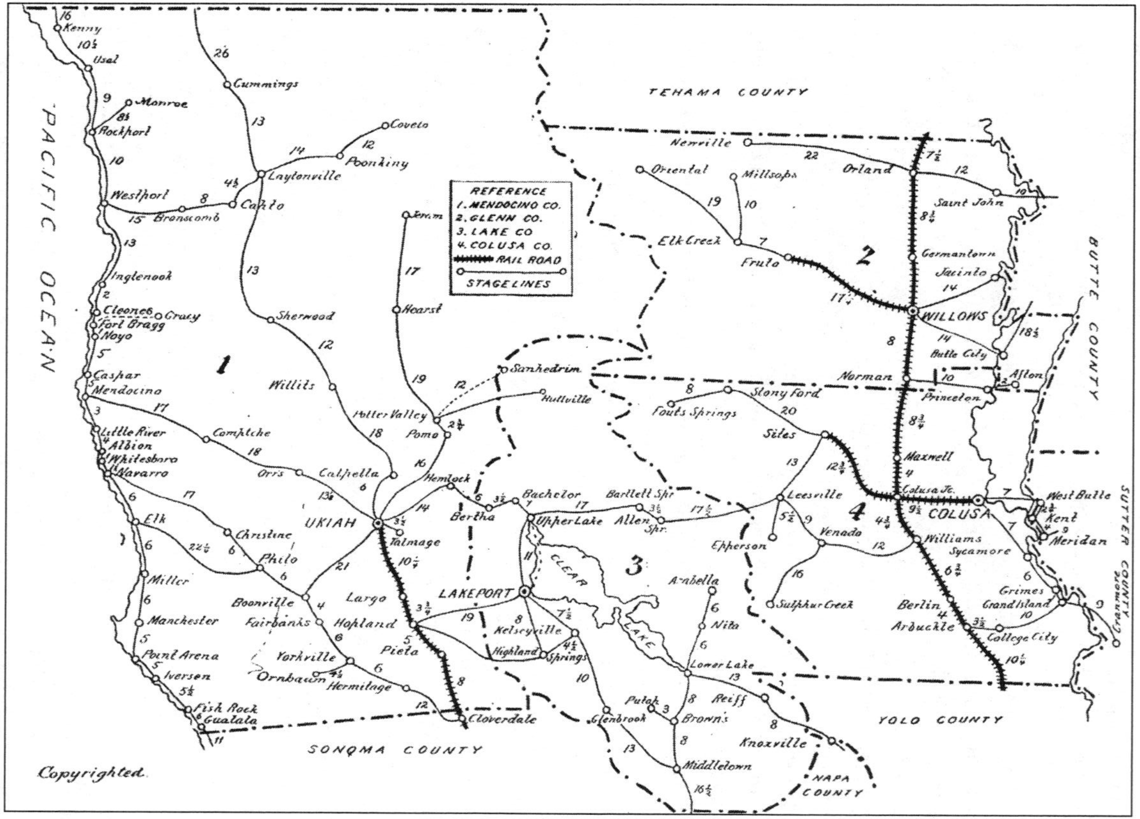

FIGURE 9.2. Map of stage and railroad lines in Mendocino, Glenn, Lake, and Colusa counties, ca. 1900. Courtesy of the Lakeport Historic Courthouse Museum, Lakeport, California.

sold fresh, compared with only 17.4 percent of pears from the state as a whole. The county's better accessibility thus allowed it to sell fresh pears profitably to populous centers to the south, and the county's agriculture flourished for years (Stollsteimer et al. 1975, 3). In recent years, however, many pear orchards have been replaced by vineyards and wineries.

TOLL ROADS AND STAGE COACHES

After Lake County became a separate entity, residents realized that they needed better ways to get in and out. Thus began the period of toll roads. Three initial roads were constructed for different directions out of the county. The Dodson Toll ran between Lakeport and Cloverdale, where the railroad ended at that time. Built in 1865, it was 24 miles long and was also called Hard Scramble or the Old Cloverdale Road. The Lawley Toll Road was constructed in 1867 between Lake County and the Napa Valley to the south from Middletown (Lake County) to Calistoga (Napa County) over the Mt. St. Helena mentioned in Stevenson's writings. This was a very useful and popular route for travelers to and from the Bay Area. The third toll road, Blue Lakes, was built from Upper Lake to Ukiah, also in 1867. This route was less used by nonlocal travelers, because Ukiah was not connected to the railroad until 1888. Toll roads were built, maintained (not always at a high level), and operated by private concerns for many years. In essence, most people bound for Lake County or attempting to leave had to pay if they wanted to travel on a halfway passable route.

Many other toll roads followed. These included the Bartlett Springs Toll Road, the Long Toll Road, and the Highland Springs Toll Road. The Bartlett Springs road was built in 1872 and 1873 and became the connection from Lake County to Colusa County to the east. In 1890, the Long Toll Road was completed by Colonel Fred Long. It offered a shorter way from Lakeport to the railroad and ran to Hopland, where the rail line had extended to from Cloverdale to the south. The Highland Springs and Squaw Rock Toll Road ran from south of Clear Lake to Pieta, near Hopland and was constructed in 1891. It was a longer route than the Long Toll road, but it became a more popular option apparently because of its easier grades, which made it a better freight route (Mauldin n.d., 738). Times changed, though. For example, a new road built in 1896 between Upper Lake and Ukiah along a similar route as the Blue Lakes Toll Road supplanted the older, more established roadway. This route became the first public, nontoll roadway out of the county (fig. 9.2; Feltman 1993, 118; Palmer [1881] 1995, 145).

The opening of stagecoach routes that followed the various toll roads was the natural result of limited access to the county. Stagecoaches provided dependable transportation of people and goods to and from the county following regular schedules. For example, in 1881 A. J. Gibson was operating a line from Ukiah via Blue Lakes, Pearson Springs, and Witter Springs to Upper Lake. W. F. Fisher's stage company ran between towns on Clear Lake and the train at Calistoga via the rugged Cobb Valley route at that time as well (Palmer [1881] 1995, 183). By 1880, six Wells Fargo offices had been established in the county to help deliver mail and other goods to and from the area (Beck and Haase 1974, 51).

Heavier freight haulers also found lucrative work moving goods in and out of the county. One gentleman, Jake Keithly, hauled goods over the Mayacamas Mountains on the various toll and free roads for fifty-three years (1885–1938), first with wagons, and then, beginning in 1915, with motor trucks. In the early days, journeys took two to three days round trip. For isolated Lake County, the services these transporters and stage line operators provided were invaluable.

The following colorful account of a trip by a young lady from San Francisco to Lake County's Seigler Springs Resort in 1885 illustrates the novelty and adventure of traveling such a distance in one day:

> We leave San Francisco early in the morning and take the ferry-boat to Vallejo. It is by far the most pleasant part of our journey. At Vallejo we catch the train and journey more or less northward to Calistoga. Train travel leaves much to be desired. If we open the window cinders blow in and if we leave it closed the air becomes appallingly close. There is a group of men in one corner playing cards and occasionally making the spittoons to ring with their wads of tobacco. When the porter comes through with a basket of refreshments that may be purchased they practically clean him out of the miniature whiskey bottles. By the time we reach Calistoga they are quite tipsy. Aunt Hester says they should all be thrashed.
>
> The stage is waiting for us in the little town of Calistoga. It is 16 feet long and 12 feet high, so covered with dust that one cannot tell if it has been painted or left with varnish over the original wood. The rear wheels are 6 feet high and tower over our skinny driver; the front wheels are smaller, being only around 4 feet in diameter. There are 3 span of horses.
>
> We board the stage while the driver and his helper store our luggage in the "boot." Our round-trip fare has cost us each $7.00. Travel has become simply shocking. Dear knows where it will end!

> The driver climbs on board and the ladies all adjust their dusters. The passengers brace themselves. "Gee-haw!" the driver shouts and the stage lurches forward, throwing us all into each other's laps.
>
> The road is narrow and winding, twisting tortuously up the narrow mountain gorges and down the steep canyons. Dust boils up around the stage. Its leather thongs, which are its only springs do their best, but we bounce on every chuckhole and occasionally fly upward to bump our heads on the roof when the big wooden wheels strike a "hard-head" [protruding rock?].
>
> There are times when the wheels of the coach go dangerously near to the edge of a cliff and the whole body sways so as to give the impression we are sure to go over. Coming around a narrow curve we almost collide with a buggy drawn by a single horse. It is such a close call that the buggy is forced to drive so that the wheel is partially up the side of the cliff. His horse slips and goes down, entangled in the reins.
>
> For a few moments the driver of the stage and the owner of the buggy shout at each other and it looks as though violence is imminent, but then they help the horse to its feet, untangled the reins, agree that the roads hereabout [are] murderous, but they wouldn't live in the city if it was give to them, and we are once again on our way.
>
> One of the gentlemen passengers explains that we must not be concerned for this is one of the best drivers on the route and one of the most experienced teams. The two rear animals are chosen for their size, being larger than the ones in front. On sharp turns the driver allows the horses to deal with the situation and the leaders pick up speed, swinging out, while the wheelers hold back just enough to keep the stage under control.
>
> Be that as it may, Aunt Hester has become ill with motion sickness and must take a sip from the medicine bottle she carries in her handbag. (M. Johnson 1977, 8–11)

The toll roads and stagecoach lines were the primary methods of travel to and from the county until about 1906, when the horse stages were replaced by auto stages. Many of the toll roads continued their operation, now charging for motor vehicles as well as animal-powered transport. Gradually they began to transition to free use under public ownership. A sign on the Bartlett Springs Toll Road from Clear Lake to Bartlett Springs in 1918 indicated the one-way toll for a six-horse team as $1.00, a four-horse team at $0.75, a motor vehicle at $1.25, motor cycles at $0.50, and each head of livestock at $0.07 (Feltman

FIGURE 9.3. Stagecoaches in front of Mt. St. Helena Inn, Napa County. Many travelers stopped here to and from Lake County, date unknown. Courtesy of the Lakeport Historic Courthouse Museum, Lakeport, California.

1993, 117). By the early 1920s, "free" roads and highways finally became the rule rather than the exception (fig. 9.3).

FURTHER ECONOMIC DEVELOPMENT

With the continued construction and improvement to the toll road and stagecoach system, Lake County was ready for further growth. Agriculture continued its relative importance but was not the only source of employment. Manufacturing was clearly not the best economic option, as the physical barriers to transportation and the distances to major population centers canceled any advantages the area might have had for producing goods. The forest industry, a mainstay of the northern California coastal counties, was not unheard of, but its value appears to be mostly in production for local purposes. The *History of Lake County 1881* gives statistics for lumber production in 1868 as 1.7 million board feet, 1873 as 3.5 million board feet, and 1880 as 1.0 million board feet (Palmer [1881] 1995, 88). The manufacturing and lumber industries never became very important, but quicksilver (mercury) mining did develop into a mainstay during the late 1800s, which was a pattern found in much of California and the West.

Quicksilver prospecting began in the early 1860s in Lake County, but significant amounts of this strange liquid metal were not produced until 1873, when the Great Western mine started to operate. The Great Western was located within the Mayacamas (or Mayacmas) quicksilver district, one of four important quicksilver areas in the county. This particular district covered the southwestern part of Lake County and extended into Sonoma and Napa counties. The other three zones of quicksilver-bearing geologic material were Clear Lake, around the southern part of the lake; Knoxville, mostly in Napa County but also spreading into southeastern Lake County and a portion of Yolo County; and Sulphur Creek, east of Clear Lake astride the line between Lake and Colusa counties (Bradley 1918, 52).

These districts were all in locations remote from county population centers, except Clear Lake. For example, travel to the Mayacamas district was via the southern toll road from Calistoga. On this road, "there was a long and steep grade on either side of the summit. Thus, whether entering or leaving Lake County, there was always a grueling pull for the endless succession of horses which, in the days before the automobile came to their relief, toiled and sweated and sometimes dropped dead in the traces as they hauled the heavily laden freight wagons or the sardine-packed stages 'over the mountain'" (Goss 1958, 5).

Once the criminal element caught wind of the large amounts of mining money being transported along the roads, their attention also turned to the Wells Fargo stage, which delivered money for distribution to the miners.

> The stages plying between Calistoga and Middletown were robbed fairly frequently, and since the robbers schemed to pull off their hold-ups on days when the strong box contained rather large sums for the mines, sometimes the Great Western [mine] money was included in the bandit's haul. The loss, of course, did not fall on the company, since Wells Fargo made it good; but even with the low pay scale of those days, the amount was several thousands of dollars when the crew was at full strength. (Goss 1958, 51–52)

Despite the rough roads and the prospects of a stagecoach holdup, the allure of natural resource profits brought many laborers into the county for a time. The Sulphur Bank Mine, near the shores of Clear Lake, emerged from the remains of the short-lived California Borax Company operation, which had extracted boric acid, sodium, and then sulphur from the site between 1864 and 1870. Quicksilver mining grew rapidly in the 1870s. At one point, there were upward of 1,000 residents at the Sulphur Bank Mine, 600 of whom were Chinese laborers. The Great Western mine operated from 1873 until 1909. Here also, much of the work was done by Chinese, up to 250 at one time, although

FIGURE 9.4. Postcard showing a sailboat and a steamboat on Clear Lake. Mt. Konocti is in the background. Courtesy of the Lakeport Historic Courthouse Museum, Lakeport, California.

Italians replaced many of them by the early 1900s. The mines helped double the population of the county from 2,969 in 1870 to 6,596 by 1880.

The ironic thing about travel in Lake County was that once one reached Clear Lake itself, travel time could be greatly reduced and comfort improved by continuing on the water. The recreational value of boating on Clear Lake was almost immediately recognized, and by the early 1860s sailing vessels were being constructed to take pleasure seekers on the waters (fig 9.4). In 1873, the first steamer, "Hallie," was launched and thus began an important period of lake transport, with many other steamers plying Clear Lake over the years. Boat service between lakeside communities became very important, and all of the county's most substantial towns, except Middletown, developed near the water's edge.

> Besides pleasure, several of these boats were used for mail delivery, passenger service, and freighting, making regular runs between Lower Lake, East Lake, Soda Bay, Paradise Valley, Lucerne, and Upper Lake, and for a number of years the steamers were the only link between several of these communities and the outside, since at this time there were no highways connecting the entire county with each community. (Geoble 1972, 1–8)

To this list of communities that benefited from lake transport must be added Lakeport, which became the county seat and was very important as a center of boat travel. The agricultural areas and mining towns of Lake County prospered throughout much of the late nineteenth century. Concurrent development of toll roads and boat service on the lake allowed growth to continue and flourish despite Lake County's remote setting. Specifically, by 1890, the county's population had increased to 7,101 from 2,969 in 1870. This was the highest census total of county inhabitants until 1930.

RESORTS

Throughout the late 1800s into the early twentieth century, another economic resource helped attract people and money to Lake County: hot and cold springs resorts. This industry developed into an important component of the county's economy, long before other western counties came to depend on tourism. The story is not unique at first glance, since many mineral spring resorts also sprang up in neighboring Sonoma, Napa, and Mendocino counties, as well as across the country. However, the concentration of springs and their success despite the extra travel effort required to access them, testify to their particular importance.

The many springs were well known to Native Americans who had lived around Clear Lake. However, in 1852, Archibald Ritchie "discovered" some mineral springs near what became Middletown. Mineral springs resorts began to be established in other areas of the state in the 1850s and 1860s, the first of which, White Sulphur Springs, opened in Napa County near Saint Helena in 1852 (Young 1998, 8, 36). It wasn't until around 1870 that such resorts began in Lake County, but Ritchie's site, Harbin Springs, was one of the first. It opened in 1871 and was the first of many (Young 1998).

Adams, Allen, Bartlett, Highland, Howard, Saratoga, Seigler, and Witter were among the more notable resorts that sprang up around the county in the 1870s. By the mid-1890s, Lake County boasted some 10 percent of all the mineral springs in California, as analyzed by Winslow Anderson for their chemical content and curative abilities (Anderson 1890). Such locations were places where people would vacation for weeks at a time, which meant vacationers were more willing to endure some extra travel. Advertising promoted the fact that with a little effort the resorts could be reached in one day's hard travel from San Francisco, even though one might be very frazzled by the end of the trip, as described earlier (figs. 9.5, 9.6).

The resorts and Lake County did all that they could do to promote themselves. Claims abounded that the waters could cure people of diseases from

FIGURE 9.5. Drinking the healthy waters of Bartlett Springs, date unknown. Courtesy of the Lakeport Historic Courthouse Museum, Lakeport, California.

FIGURE 9.6. Witter Springs Resort, northwest of Upper Lake. This photo shows the impressive size of some of the major resorts in the county. The building was completed in 1906 and had 100 rooms, 500 windows, 20 miles of electrical wiring, and cost over $250,000 (Feltman 1993, 102). Courtesy of the Lakeport Historic Courthouse Museum, Lakeport, California.

rheumatism to gout, diabetes to constipation (fig. 9.7). "Lake County contains more mineral springs than the whole of Europe, and the day is not far distant when invalids from all climes will resort here for health and pleasure," states a booklet on the county (Board of Supervisors [1888] 1988, 13). Winslow Anderson's book on California's springs describes Harbin Springs.

> It would be difficult to find a more delightful or picturesque spot so rich in scenery, mountains perpetually clad in evergreen garments and hills with verdure bright forming the background of the landscape.... The climate, too, is one of the most salubrious in the State. (Anderson 1890, 164, 166)

Additionally, Highland Springs reminded everyone in one of their advertisements that Lake County was the "Switzerland of America" (Board of Supervisors [1888] 1988; Pierucci 2016). An old postcard in the Lakeport Historic Courthouse Museum shows how Lakeport played on this Swiss association, billing itself "The Geneva of America."

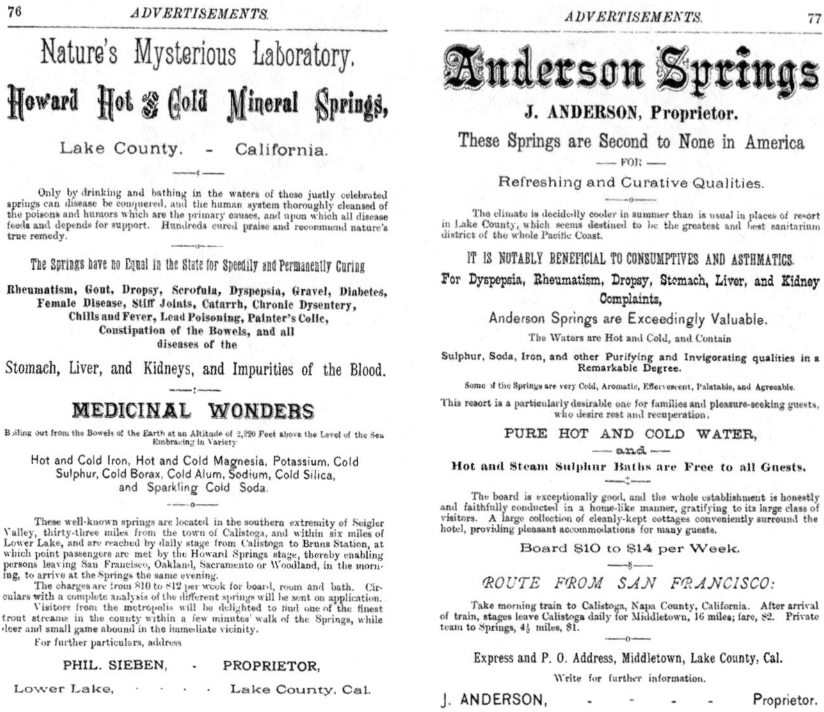

76 *ADVERTISEMENTS.*

Nature's Mysterious Laboratory.

Howard Hot & Cold Mineral Springs,

Lake County. - California.

Only by drinking and bathing in the waters of these justly celebrated springs can disease be conquered, and the human system thoroughly cleansed of the poisons and humors which are the primary causes, and upon which all disease feeds and depends for support. Hundreds cured praise and recommend nature's true remedy.

The Springs have no Equal in the State for Speedily and Permanently Curing

Rheumatism, Gout, Dropsy, Scrofula, Dyspepsia, Gravel, Diabetes, Female Disease, Stiff Joints, Catarrh, Chronic Dysentery, Chills and Fever, Lead Poisoning, Painter's Colic, Constipation of the Bowels, and all diseases of the

Stomach, Liver, and Kidneys, and Impurities of the Blood.

MEDICINAL WONDERS

Boiling out from the Bowels of the Earth at an Altitude of 2,220 Feet above the Level of the Sea Embracing in Variety

Hot and Cold Iron, Hot and Cold Magnesia, Potassium, Cold Sulphur, Cold Borax, Cold Alum, Sodium, Cold Silica, and Sparkling Cold Soda.

These well-known springs are located in the southern extremity of Seigler Valley, thirty-three miles from the town of Calistoga, and within six miles of Lower Lake, and are reached by daily stage from Calistoga to Bruns Station, at which point passengers are met by the Howard Springs stage, thereby enabling persons leaving San Francisco, Oakland, Sacramento or Woodland, in the morning, to arrive at the Springs the same evening.

The charges are from $10 to $12 per week for board, room and bath. Circulars with a complete analysis of the different springs will be sent on application.

Visitors from the metropolis will be delighted to find one of the finest trout streams in the county within a few minutes' walk of the Springs, while deer and small game abound in the immediate vicinity.

For further particulars, address

PHIL. SIEBEN, - PROPRIETOR,

Lower Lake, - - - - Lake County, Cal.

ADVERTISEMENTS. 77

Anderson Springs

J. ANDERSON, Proprietor.

These Springs are Second to None in America

— FOR —

Refreshing and Curative Qualities.

The climate is decidedly cooler in summer than is usual in places of resort in Lake County, which seems destined to be the greatest and best sanitarium district of the whole Pacific Coast.

IT IS NOTABLY BENEFICIAL TO CONSUMPTIVES AND ASTHMATICS.

For Dyspepsia, Rheumatism, Dropsy, Stomach, Liver, and Kidney Complaints,

Anderson Springs are Exceedingly Valuable.

The Waters are Hot and Cold, and Contain

Sulphur, Soda, Iron, and other Purifying and Invigorating qualities in a Remarkable Degree.

Some of the Springs are very Cold, Aromatic, Effervescent, Palatable, and Agreeable.

This resort is a particularly desirable one for families and pleasure-seeking guests, who desire rest and recuperation.

PURE HOT AND COLD WATER,

— and —

Hot and Steam Sulphur Baths are Free to all Guests.

The board is exceptionally good, and the whole establishment is honestly and faithfully conducted in a home-like manner, gratifying to its large class of visitors. A large collection of cleanly-kept cottages conveniently surround the hotel, providing pleasant accommodations for many guests.

Board $10 to $14 per Week.

ROUTE FROM SAN FRANCISCO:

Take morning train to Calistoga, Napa County, California. After arrival of train, stages leave Calistoga daily for Middletown, 16 miles; fare, $2. Private team to Springs, 4½ miles, $1.

Express and P. O. Address, Middletown, Lake County, Cal.

Write for further information.

J. ANDERSON, - - - - Proprietor.

FIGURE 9.7. Advertisement for Howard Hot and Cold Springs and Anderson Springs, 1888. Board of Supervisors of Lake County 1988, 76–77.

The trip to the county and springs was also romanticized. "The mountain roads are well kept, and the stage ride is one of the most picturesque in the State" (Anderson 1890, 82). "The roads are generally in good order and make access easy to all parts of the county; where they cross the mountains the picture is ever-changing, here we see a beautiful, smooth, symmetrical hill covered with stately pines or noble oaks" (Board of Supervisors [1888] 1988, 14). Compare these statements with those of the young traveler above! Apparently, those writing resort advertising were as willing as modern advertisers to overlook any negative aspects of their destinations.

Thousands heeded these siren calls to Lake County and made the trip there. Even without a railroad, the county's proximity to bustling San Francisco gave it an advantage over more far-flung resorts. An 1870 census counted some 5,963 tourists in the county, nearly twice the number of permanent county residents enumerated by the official census (Smith and Broderson 1989). Later, in 1883, it was estimated that between 15,000 and 25,000 people visited the county each year "in search of health and pleasure" in spite of the transport challenge

(Mauldin n.d., 1765). After the turn of the century, many of these resorts slowly died, losing popularity due to changes in attitudes regarding the supposed health benefits of hot springs, the shift to beach tourism at Clear Lake, and the rise of mountain resorts in the south part of the county. Towns referencing exotic cities such as Nice (France) and Lucerne (Switzerland) sprang up on the lake in the 1920s, and mountain destinations such as Hoberg's Resort in the pines near Cobb replaced the springs resorts as vacation destinations.

THE RAILROAD THAT WASN'T

For decades, a railroad connection for Lake County to the larger world was seen as the key to development. It never came. The story of this long-anticipated link was lengthy and full of disappointments for county residents during the late 1800s and early 1900s. The writer of *History of Lake County 1881* summed up the state of affairs.

> A railroad is the greatest immediate want of Lake County; but we are sorry to record the fact that the prospect of getting an outlet of this kind in the near future is anything but flattering. Ever and anon excitements about railroads rise to fever heat, and the people can almost imagine that they hear the shrill whistle of the engine re-echoing through the valleys and mountains; but the interest subsides, and all is as quiet as the grave on the subject. (Palmer [1881] 1995, 94–95)

The year 1881 was not the end of this dream. By 1888, county leaders were sure that the railroad was really on its way.

> The day, however, is not far distant when Lake will be connected with other parts of the State by railroad, two of which are now projected and the completion of one of which, to connect Napa with the waters of Lower Lake, is promised within the present year. (Board of Supervisors [1888] 1988, 15)

Numerous promoters continued to push different routes but always came up short. A 1910 plan drew a course right through the mountains using two tunnels, one 700 feet and the other 4,000 feet, at an estimated cost of $376,000 to get from Santa Rosa to Lakeport (Cottrell 1910, 16). Initial grading and development of part of the route from Hopland, Mendocino County, was even started in late 1911. These grand ideas stayed alive until around 1916, but sufficient financing never came (Pierucci 2015). By then the golden period of railroad construction had passed (Simoons 1953). So the closest railroad lines

FIGURE 9.8. High Hopes for a railroad coming to Lake County, ca. 1909. This event never occurred. Courtesy of the Lakeport Historic Courthouse Museum, Lakeport, California.

continued to be where they had been for years (Calistoga, Napa County, to the south; Ukiah, Mendocino County, to the west; and Sites, Colusa County, to the east), the starting points in other counties for the horse and auto stages that entered the rugged hills, interspersed valleys, and primary towns of Lake County (fig. 9.8).

Although Lake County may have been constrained by the railroad that wasn't, the advent of the automobile and the construction of paved highways and county roads proved to be the most beneficial transportation improvements. Californians began to flock to the area, starting in the 1940s, when county population increased from 8,069 in 1940 to 13,680 by 1950. The county had been rediscovered as a vacation spot, and now paved roads allowed the lake and mountain resorts to prosper. Retirees were also attracted to the region and became important contributors to the economy (Nixon 1966).

FROM REMOTE FRONTIER TO VACATIONLAND

Lake County has been "discovered" many times over its history. Each discovery has been complemented by transportation improvements that allowed the area to grow in population and more distant connections. Mountain men interacted with the Native Americans, making their way on trails. Frontiersmen displaced the Indians and opened roads to bring in yet more people. Farmers harvested crops from the fertile soils of the low valleys, while miners and opportunists opened the remote hillsides to quicksilver extraction, and entrepreneurs established mineral spring and other resorts. Toll roads and stage lines began operating to meet the demand of these economic activities, while regularly scheduled water transportation accelerated interactions among the communities near Clear Lake. As mining and mineral springs resorts declined, highways and automobiles brought people in faster, and the lake in the middle of it all began to receive more attention. The county survived and even prospered.

The potential for profit never seemed great enough to justify the cost of building a railroad over the mountains to the county. Even today, the possibility of large population increases is unlikely, despite county proximity to the Bay Area. However, the county has never been doomed to failure. In reality, Lake County's variety of resources, especially its recreational and environmental attractions, have helped to outweigh its winding, difficult roads. The next chapter expands the scale of this historical settlement exploration by incorporating two additional counties, Mendocino and Sonoma, into an analysis of how this region of north coast California developed differentially according to local geographic and socioeconomic circumstances.

Lake County now has a greater population density than Mendocino County to the west and Colusa County to the east. Both neighboring counties had railroad connections over a century ago, but they have not had the same types of economic activities and attractions that have allowed Lake County to grow despite its ring of physical barriers. One can still get motion sickness driving Lake County, but people make the trip for the same reasons they always have—the destination is worth the effort.

10

Far-Off Frontiers Emerge as Fine Farms and Towns North of the Bay

Our earlier acquaintance with the Winklers and Greenwoods, pioneer settlers of Sonoma and Mendocino counties, helps set the scene for the region's rapid growth. Clayton Winkler was just one farmer, and adding his boys made some more. They were part of the larger economic scene that composed Sonoma County. Each individual shaped the whole; each county market would not exist without the individuals who made it run. Clayton was one of the fortunate ones who escaped the often-disillusioning life at the mines and made a new life near the western sea. Hundreds, even thousands of others like him made their way to the region and built a thriving agricultural center.

A similar thing can be said for the Greenwoods in Mendocino County. They settled in a less agriculturally promising place but one that had tremendous timber and sea resources. Brothers James, Boggs, William, and Britton Greenwood were among the first to settle Cuffey's Cove on the coast in 1854. "They lived in a square house constructed of split redwood boards, which were riven on the top of one of the adjacent hills and brought down to the building site on the back of a horse. This was doubtless the first real house that was ever built in the township [of Big River]" (Palmer 1880, 400). The fascinating story of how the early settlers like the Winklers and Greenwoods transformed a promising yet quiet corner of California into an integrated spatial system in record time is the subject of this chapter (Otterstrom 2007).

METHODS AND DATA: THE REGIONAL SETTLEMENT PROCESS

Three northern California counties of Sonoma, Mendocino, and Lake are excellent sites for studying how a regional system develops beyond the core city. Historically, the area has been set between the urban center of San Francisco and the resource-dependent periphery of far northern California (fig. 1.1). Both core and periphery characteristics have been evident in these counties during different periods, together making them an appropriate case study for understanding the differences between urban and rural settlement structure and economic transformations over time. The counties were not known for their gold deposits and were essentially a secondary destination for settlers, but because they were in the shadow of San Francisco, they still experienced steady and often rapid settlement.

Additionally, this example also permits study of the process without having to recreate long settlement histories. The influx of settlers translated into the compressed formation of a regional structure with hierarchically differentiated cities. The good census and county directory data available for the years in question make this area easier to study than places in the eastern United States that were settled much earlier. Finally, even though the region probably experienced more rapid development than many other parts of the country, it can serve as a test case to illustrate the processes of settlement diffusion, which can then be applied to other places and times in American settlement historical geography.

My analysis of the three counties is divided into two parts: (1) topographical diversity and (2) population structure from 1850 to 1880. The topographical diversity section backgrounds the geographic situation and the importance of its variety of terrain and vegetation. It highlights the settlement constraints and opportunities associated with the coastal areas, valleys, and mountains of these counties. The geographic background forms the basis for the population growth and economic structure that developed in the region.

The population section considers changes in the settlement landscape between 1850 and 1880. Discussion includes a short analysis of the overall settlement development, or diffusion, in the three study counties plus reference to the influence of San Francisco County. I place the population shifts within the spatial context of their diversifying economies by analyzing the counties' changing occupational structure. This section also shows that as the new region grew, its settlement hierarchy not only diversified but also became more stable in form and function.

County Directories and Censuses

The prime data sources for the population analysis were *L. L. Paulson's Hand-book and Directory of Napa, Lake, Sonoma, and Mendocino Counties* (1874), samples of households from the U.S. censuses from 1850 to 1880, and the 1852 California Census for Sonoma County (Harris 1983, 1992. See also Conzen 1972; Conzen and Conzen 1979; Hardwick 1990). *Paulson's Hand-book* includes a list of adult citizens, their occupations, place of business, residences, and names of their local post offices. It itemizes some businesses as well. Although the *Hand-book* does not include all of the counties' adults, this directory has one advantage over the official U.S. Censuses: it gives the place of residence more specifically than the U.S. Census, enabling location on a map. This allows illustration of the economic pattern of each county both visually and statistically, making the directory an excellent addition to the basic decennial census records.

Thus, an important component of the research was mapping the *Paulson's Hand-book* directory listings for Mendocino, Sonoma, and Lake counties. The residence locations were originally written in the directory using cardinal directions and distance from the nearest town (e.g., 3 miles northwest of Santa Rosa). If a person lived in a town, the street address was given. After each of the entries was entered into a database, I converted the listed locations of each residence to latitude and longitude coordinates. These coordinates became the basic data for mapping occupations.

Getting reliable map coordinates presented several challenges. First, a number of the listed towns no longer exist and so were difficult or impossible to find. Forgotten places like Hermitage and Glenmark in Mendocino County and Clairville in Sonoma County have disappeared from most modern maps (see fig. 10.1 for the locations of many of the towns mentioned in *Paulson's Hand-book*). Second, the directory lists numerous residences at a single spot (e.g., 9 miles north of Ukiah), so many original mapped points overlapped. I addressed this problem either by using graduated circles for locations with multiple entries, or, in the case of some of the maps, by offsetting multiple entries slightly from their listed locations.

I made maps to compare the types and numbers of occupations by county to develop a structure of the economy from core to periphery by industry. The core in any settlement system is the area of greater population density and hence greater economic influence. Farther out, the areas that have lower population densities, less economic power, and that rely on services from the core are known as the periphery. This idea of a core/periphery structure has been used from the world scale (e.g., Wallerstein 2004) down to more regional scales (for a Chinese example see Skinner [1977]). The conceptual framework is important here because it helps explain the economic patterns between the core in Sonoma County and the more peripheral areas of Lake and Mendocino counties.

After locating the directory entries for each listing, I plotted groups of

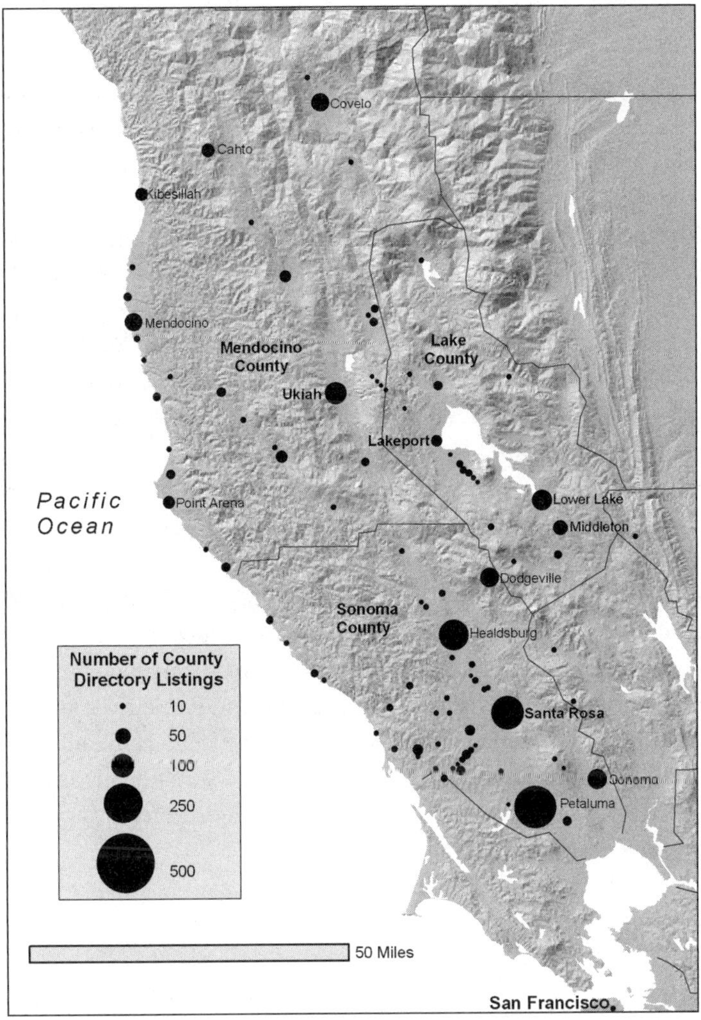

FIGURE 10.1. Listings from *L. L. Paulson's Hand-book and Directory of Napa, Lake, Sonoma, and Mendocino Counties* (1874). Spencer Larson, BYU Geography.

people according to their occupation. For the purpose of this chapter, I first mapped all the points to show the basic spatial density of the county. I then made specific maps for various types of occupations to determine any patterns. The exercise proved valuable as a way to better show the place-specific distributions of the region's citizens and to underscore the hierarchical structure of economic differentiation.

ASSESSING TOPOGRAPHICAL DIVERSITY

Migrants settled Sonoma, Mendocino, and Lake counties soon after San Francisco. Before the 1850s, the Euro-American population was very sparse in the region. The bulk of the new settlers found their way to Sonoma County, while the Lake and Mendocino areas were virtually undiscovered until after 1850. The land that became Sonoma County was the most accessible and included the northernmost Spanish mission (San Francisco Solano, established in 1823) at what now is the town of Sonoma. Additionally, the southern reach of the Russian fur trade was along the coast in future Sonoma County in the vicinity of Fort Ross, which was in use between 1812 and 1841 (Gudde 1949, 119; H. Hansen and Miller 1962, 15–19). Sonoma County's early communities included Sonoma (the town), Petaluma, and Santa Rosa, established in 1835, 1851, and 1851 respectively (Gudde 1949). The 1850 census reported 560 people in Sonoma County, 55 in Mendocino County (mostly men), and only a few families in the area that would become Lake County in 1861.

Although the topographic base does not determine the unfolding of the system of inhabited places, it does provide a distinct stage on which these activities proceed. Three main divisions of the region's physical geography are recognizable: the coast, the coastal mountain ranges, and fertile river and lake valleys in the interior. The first two divisions saw fewer settlers initially, while the valleys were the prime destination for new arrivals. Additionally, even though the climate varied throughout the region, it is and was generally agreeable and was therefore a selling point for new settlers.

Coast

The northern coast of California was rugged and in many places nearly impassable. Without exception, ports along the Sonoma and Mendocino coasts were small, possibly because of the rugged hills that rose directly to the east. The outlets of several rivers (e.g., Russian and Noyo rivers, which empty at Jenner and Fort Bragg respectively) and good natural harbors (such as Bodega Bay) proved to be the most promising coastal locations, but these still did not grow substantially. This is true even though the redwood logging of the late 1800s boosted coastal shipping for a time. While the inland reach of sea trade was not far, Sonoma County was an exception. On the southeast, it adjoined the more accessible San Pablo Bay, which in turn neighbored San Francisco Bay. The transportation advantage was vital for Sonoma County's future success.

Mountains

Settling in the coastal mountain ranges was not an easy task. The forest was thick, slopes were extreme, and newcomers had to carve out homesites on the most level spot available. Sonoma County was the least rugged of the three counties, but the coastal range in this area was still difficult to traverse in many places. For even more variety in elevation, one had only to look to Lake and Mendocino counties. Mountains ringed Lake County, which made it particularly difficult to reach the Clear Lake area and delayed settlement until good roads were constructed (Simoons 1953; chapter 9, this volume). Mendocino County was equally rugged. The resulting uneven population distribution across the counties simply attested that more pioneers chose not to homestead and farm in the mountains (fig. 10.1).

The three counties fell in a relatively medium range of California ruggedness and accessibility. On one side, the hills of Lake County overlooked the flat and fertile Sacramento Valley, which struck a strong contrast between two topography types. On the other hand, the Sonoma, Lake, and Mendocino study region was less mountainous than the Humboldt, Trinity, and Del Norte country farther north. Additionally, Sonoma, Mendocino, and Lake counties had another important advantage over the far northern counties: they were substantially closer to San Francisco, which invited more economic interaction and physical linkages.

Yet although the mountains generally saw fewer settlers, some economic activities were especially successful there. Logging, ranching, quicksilver mining (silver and quicksilver mining were concentrated in the mountains of southern Lake and eastern Sonoma counties for a time), and mineral springs resorts (especially in Lake County) were very successful in their day (Palmer [1881] 1995). In this terrain, redwood logging was most noteworthy as a population lure. Just inland from the coasts from northern Sonoma County through Mendocino County and beyond, the thick redwood forests were cut at an amazing rate beginning in the 1850s. *Paulson's Hand-book* of 1874 showed that most of the lumber workers were found in these northern Sonoma and Mendocino areas. A large number of mill hands were listed on the central coast at the Big River mill near Mendocino City, the J. G. Jackson mill in Caspar, and the Coombs and Perkins mill near Mendocino/Little River. Most of the employees of the other working mills did not make it into *Paulson's Hand-book*, even though there were many mills operating throughout the area (Palmer 1880, 141). All in all, the coastal ranges provided income to new settlers, but it was hard fought.

Valleys

Even though the coastal and mountain regions had their appeal, early explorers found the flatter valley lands more inviting: moisture was adequate; temperatures were moderate; and the soils proved to be productive. When the good word reached people in the eastern United States, as well as successful and unsuccessful Sierra miner alike, these northern California counties attracted a great number of new settlers.

Some specific valleys deserve attention. The lowlands along the Russian River were a prime settlement destination in southern Mendocino and northern Sonoma counties, with the notable towns of Ukiah, Healdsburg, and Cloverdale developing along its banks. In Lake County, flat land was limited to the area around Clear Lake—known as Big Valley—and smaller valleys between the rolling hills and mountains (e.g., Scotts Valley, Long Valley, and Clover Valley). Round and Potter valleys of Mendocino County were harder to reach but also had promising farmland.

Southern Sonoma County had particular advantages. It was large, flat, and closest to the San Francisco area. Petaluma, Santa Rosa, and Sonoma grew with the farms in the surrounding countryside that were nurtured by abundant sunshine, sufficient moisture, good soils, and comfortable temperatures. Most significantly, growers soon found the area was an excellent environment for vineyards and wine production. By 1874, population was relatively dense in these southern towns and surrounding farmland.

POPULATION ANALYSIS: DIFFUSION, DIVERSIFICATION, AND STABILITY

Three interrelated issues are addressed here. The first is the founding and growth of new settlements. The second is economic diversification, specialization, and the resulting differential growth of the communities within the regional structure. Finally, there is the evolution of the region toward settlement structure stability.

Here I argue that the spread of homesteads and towns in a new frontier, called settlement diffusion, follows the traditional diffusion process. At some point, each tract of "wilderness" will probably experience the diffusion of farms, villages, and cities across its landscape. As with other diffusion processes, this spread can occur quite quickly or can take place over centuries. At some point, diffusion saturation occurs as settlements have been founded throughout the areas of the region that are most amenable to supporting people (fig. 1.2).

During settlement diffusion there is a marked local differentiation or economic specialization and diversification among places. It occurs as people continually seek the most viable uses for their land and begin manufacturing goods or providing specialized services for local residents and those in neighboring regions. Over time, more people fill the region, but growth is unevenly distributed. People gravitate toward areas with the most potential or greatest immediate need for labor. The process is fueled by growing retail and wholesale economic relationships (as shown by Christaller [1933] 1966 and Vance 1970) and the existing physical geography or stage upon which this system of cities arises.

Settlement stability occurs when the main towns and cities in a region also become firmly differentiated in terms of relative size and function. This transformation is important because it marks the point when settlement distribution and relative size remain fairly stable. This can be seen simply by comparing relative county and city sizes over time. With hierarchical settlement stability, there may also be a measure of functional stability in terms of employment levels in the various economic sectors, but we will not track functional stability levels in this region's economy past 1880.

Settlement Diffusion, 1850–1880

Large-scale immigration into California helped speed the establishment of its settlement hierarchy. It did not take long for San Francisco to grow into one of the most densely populated counties in the nation (partly due to its small land area). San Francisco received a tremendous boost from the discovery of gold, and it became a merchandising center serving the waves of new immigrants who came to seek their fortunes. Its greatest percentage growth was during the 1860s, but every decade from 1850 to 1880 showed high rates of immigration. The city's population ballooned from a small community of about 500 in 1847 by U.S. Census accounts to over 56,000 by 1860, and nearly 150,000 in 1870 (Richards 1991, 43). By 1880, San Francisco had a population of about 234,000, with a density of 4,561 people per square mile (table 10.1). Its population dwarfed that of surrounding areas, leaving Sonoma, Mendocino, and Lake counties far behind, even though these semiperipheral areas registered significant growth themselves.

With changing transportation modes, San Francisco was able to extend its reach out. Because the city is on a peninsula, most travelers going north had to ferry across the bay to Marin, Sonoma, Napa, and Solano counties. Ferries remained the most efficient means to do so until construction of the large

TABLE 10.1. Population density per square mile and total population of Lake, Mendocino, San Francisco, and Sonoma counties, 1850–1880.

		LAKE	MENDOCINO	SAN FRANCISCO	SONOMA
1850	Density		.01		.27
	Total		55	**	560
1860	Density	N/A	1.15	1,107.25	7.83
	Total	[962]*	3,967	58,802	11,867
1870	Density	2.36	2.18	2,913.7	13.07
	Total	2,969	7,545	149,473	19,819
1880	Density	5.23	3.70	4,560.6	17.10
	Total	6,596	12,800	233,959	25,926

* Lake County did not exist until 1861. This number is the population of Clear Lake Township in Napa County, which formed much of the future Lake County.

** Data from the 1850 census of San Francisco was lost/destroyed

(U.S. Census data)

bay bridges in the twentieth century. Investors later constructed railroads up Napa Valley and through Sonoma County (by 1871) to quickly move San Francisco freight and passengers from ferry ports along the bay (Hansen and Miller 1962, 48). Some distinctive economic activities encouraged interregional relations. Direct shipping between the redwood mills along the northern Sonoma-Mendocino coast and San Francisco provided a strong tie north and south. Following Sonoma County, Mendocino County received the railroad (after 1871), but Lake County never had that transportation connection. Fortunately for Lake, its many mineral springs and resorts still attracted commerce and visitors from San Francisco, as described in the previous chapter. These transportation links further strengthened San Francisco's dominant regional status (Vance 1964).

Sonoma County quickly became the most populous unit of the three-county study region. In terms of density, Lake and Mendocino counties followed pretty closely during this thirty-year period, with some shifting in position. In 1870, Lake and Mendocino counties both barely qualified as "settled" at over two people per square mile, while Sonoma was six times more densely populated. This trend continued with Lake and Mendocino counties growing relatively quickly in terms of percentage growth, but always trailing the absolute numbers of Sonoma County. Sonoma's importance revolved around its central role in the regional lumber, mining, service, and farming industries, as well as the transportation connections that developed from within its borders to points to the north and south.

Diversification in the Economy, 1850–1880

In 1850, California had just gained statehood, and most new settlers were heading to the Sierra gold fields, the trading center of San Francisco, or places in between. Just off of this main settlement track was an area awaiting its own wave of newcomers. While Mendocino County showed its small size with all heads of households working as farmers or lumbermen, Sonoma County exhibited what appeared to be a highly diversified economy for its early age with only 56.5 percent working in the primary sector (resource extraction—mainly agriculture, logging, fishing, and mining) including laborers (table 10.2; all primary sector figures quoted here include laborers). Just two years later Sonoma County had grown substantially, but the percentage of workers in natural resource extraction had actually increased to 71.7 percent. This curious shift could have been an aberration, or it may indicate the early settlement centrality of the area, even extending back to the Mexican period.

Within both the censuses and county directories there were numerous people who reported working as "laborers." They are summarized here as working in the primary sector, because in the actual censuses it appeared that a large number of them lived in farming areas. This does not discount the fact that many others probably worked in wood milling (see for example the Laborer panel in fig. 10.4. This 1874 map shows some clustering along the Mendocino coast). Therefore, since these people could not be definitively separated out, laborers are kept separate in the tables and maps.

By 1860, distinct economic patterns among and within each of the counties are evident. Sonoma County still had a majority (63.7 percent) employed in the primary sector, but the main cities of Santa Rosa and Petaluma had already become service centers in the local settlement hierarchy, with 66.7 percent and 45.4 percent employed in services and 10.1 percent and 27.4 percent in the primary sector, respectively. Mendocino County had some 83 percent in natural resource extraction even with its steady growth. Finally, the area of Napa County that would later become Lake County seemed to be in an even earlier stage of settlement with over 90 percent working in agriculture and related activities (table 10.2).

The next two decades saw variable growth in services and manufacturing among the counties. Sonoma County's manufacturing hovered in the range of 9.3 to 12.7 percent of total employment, while its workforce share in services ranged from a low of 19.5 percent in 1870 to a high of 24.7 percent in 1880. Meanwhile, Mendocino County's manufacturing sector grew to be the largest

TABLE 10.2. Sonoma, Mendocino, and Lake county occupational structures, 1850–1880.

	PRIMARY		LABORERS		MANUFACT.		SERVICES		TRANSPORT		NOT WORKING		# OF PEOPLE IN SAMPLE	TOTAL POPULATION
	#	%	#	%	#	%	#	%	#	%	#	%		
Sonoma														
1850	105	46.7	22	9.8	39	17.3	57	25.3	2	0.9	0	0	225	560
1852	722	57.8	174	13.9	97	7.8	205	16.4	28	2.2	24	1.9	1,250	
1860	2,084	46.4	777	17.3	417	9.3	1,028	22.9	109	2.4	74	1.6	6,395	11,867
1870	199	53.8	17	4.6	47	12.7	72	19.5	6	1.6	29	7.8	370	19,819
1880	232	46.1	30	6	52	10.3	124	24.7	15	3.0	50	9.9	503	25,926
Mendocino														
1850	8	87.5	0	0	1	12.5	0	0	0	0	0	0	9	55
1860	119	71.7	19	11.4	13	7.8	11	6.6	3	1.8	1	0.1	166	3,967
1870	245	47.3	51	9.8	71	13.7	93	18	33	6.4	25	4.8	518	7545
1880	280	46.6	42	7	108	18	101	16.8	29	4.8	37	6.2	601	12,800
Lake														
1860	309	85.8	19	5.3	5	1.4	26	7.2	0	0	1	0.2	360	
(part of Napa county in 1860)														
1870	84	51.5	32	19.6	19	11.7	13	8	4	2.5	11	6.7	163	2,969
1880	90	46.4	43	17	24	9.8	28	11.2	9	3.6	30	12	225	6,596

(Source: U.S. Censuses; 1852 California Census; Harris 1983, 1992 for Sonoma County 1852, 1860)

of the three counties as its lumber mills multiplied in number and size within the redwood forests, accounting for 18 percent of those employed by 1880. Another 16.8 percent worked in services at that time. Lake County continued to be heavily reliant on the primary sector (mostly farming) with over 61 percent of the workers still employed in that category in 1880, while only 21 percent were in manufacturing and services combined.

Specialization and Occupational Structure in 1874

Paulson's Hand-book was published at a good time. It appeared in the midst of the continuing population expansion of the late nineteenth century. Opportunity continued to bring new people eager to fulfill their own California dream, which they found with varying levels of success (Burchell 1987, 177–96). Thus, the directory provides an imprint of the settlement geography at a time when the land had been thoroughly affected by European settlers, though it still had much potential for further population growth. This source also allows a large-scale comparative analysis of the economic differentiation of three largely rural counties. It thereby shows that areas relatively distant from urban centers can still display descending levels of specialization, just as large cities themselves have been shown to exhibit differentiation in various characteristics (Kish 1954, 388–98).

The contrast in numbers of occupations in the three counties is significant. The figures indicate the counties' relative levels of economic specialization. Sonoma County had 292 different occupations or types of businesses listed, while Mendocino and Lake counties had 161 and 102 respectively (table 10.3). Effort was made to combine all occupations that had different names but were essentially the same type of work. However, most probably there are still a small number of duplicates, so these occupation totals may be a little high. One point of interest is that Mendocino and Lake had relatively more occupations than Sonoma in terms of total numbers of people listed in those counties. However, Sonoma's high numbers of service and manufacturing occupations show how diversified its economy was at that early date, thus its higher order in the settlement hierarchy. Jobs there ranged through the alphabet from architect, bee raiser, and brick maker, to tinsmith, undertaker, and wool grower.

Sonoma also had the smallest percentage of workers listed within the primary economic sector (including laborers: 47.8 percent), while Lake County had the most (with laborers: 63.2 percent). Mendocino had the largest share of laborers, possibly influenced by the number working in the forests. In contrast, Lake County had the lowest percentage of people employed in the secondary or

TABLE 10.3. *L. L. Paulson's Hand-book* listings and specialization by major economic sector for Lake, Mendocino, and Sonoma counties (derived from Paulson 1874).

	LAKE	MENDOCINO	SONOMA
Primary Sector, without Laborers (% of total listings)	58.5	45.8	43.6
Laborers %	4.7	9.0	4.2
Manufacturing %	6.7	14.1	13.1
Services %	22.9	24.8	28.6
Transportation %	2.8	2.9	3.3
Not specified/Not employed %	4.4	3.4	7.2
Total Listings	879	1473	3343
Specialization by economic sector panel			
Different primary sector occupations	6	9	17
Laborers (unspecified)	1	1	1
Different manufacturing occupations	26	45	67
Different service occupations	63	95	184
Different transportation occupations	6	11	23
Total different occupations	102	161	292

manufacturing sector (6.7 percent), and Sonoma had the highest percentage of service and transportation workers (28.6 percent and 3.3 percent, respectively). Clearly, the economic landscape of the three counties had moved toward differentiation according to population size and geographic situation.

Overall population density was greatest in southern inland Sonoma County around Petaluma, Santa Rosa, and Sonoma. Northward, the numbers steadily decreased through the Russian River valley and into Mendocino County, where fewer people lived, although there were greater densities along the Mendocino coast than the Sonoma coast. In Lake County, the bulk of the population lived to the south and west of Clear Lake, as well as farther south in the Middletown region near Napa County.

The considerable variation in settlement density is further clarified by looking at the distribution of workers in several occupational fields (fig. 10.2; table 10.2). As expected, the primary sector of mostly farmers was the most dispersed and widespread, while manufacturing was concentrated in Sonoma County and along the coast of Mendocino County, where a number of sawmills were located. All three counties had a large number of service sector listings, with the bulk of these people residing in the towns. (Figs. 10.2–4 include some business operation locations that were in *Paulson's Hand-book*, in addition to individual listings, e.g., "Caldwell & Read," a dental office in Sonoma County.) There were 99 of these in Sonoma County, 26 in Mendocino County, and 7 in Lake County, with the majority in the service sector. The related statistical tables (tables 10.3, 10.5) do not include these businesses in their total figures, as they appear to duplicate individual listings).

Within the primary sector, farmers not differentiated by specialization

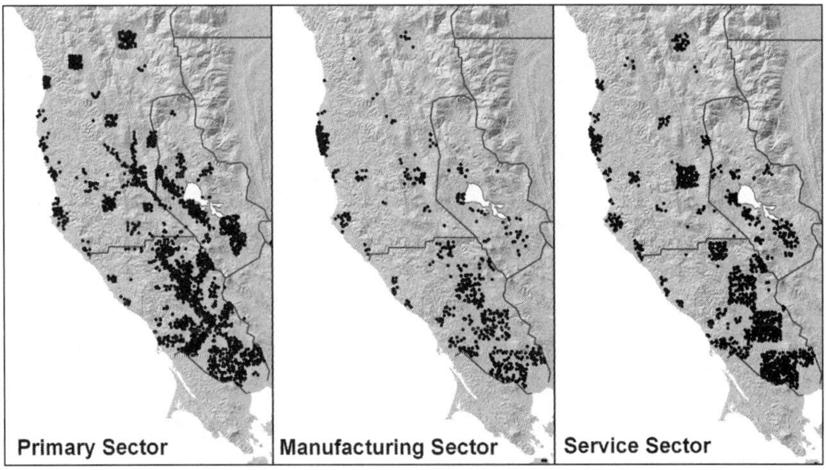

FIGURE 10.2. Primary, secondary, and service sectors (locations of listings). Data source: Paulson 1874. Spencer Larson, BYU Geography.

made up the majority of the listings, but other occupations should be noted (fig. 10.3). Dairy farmers were mostly found in southern Sonoma County near the largest population centers. Stock raisers, or ranchers, were most plentiful in northern Sonoma County as well as the Middletown (Guenoc) area of southern Lake County. The beginning of vineyards and the winemaking industry can also be seen in southern Sonoma County, with the large number of vintners there (vineyards have since spread throughout the valleys of all three counties). Finally, mining, mostly quicksilver (mercury), was concentrated in the mountainous regions of northeastern Sonoma County and the bordering areas of Lake County (Palmer [1881] 1995).

Blacksmiths and carpenters were fairly ubiquitous, while the more specialized occupations of wagon and harness makers were limited in their distribution to Sonoma County and a few select locations in the other two counties (fig. 10.3). This shows how the population differences among the three regions translated into an advantage for Sonoma County in a number of trades. The specialized occupations simply had a larger population threshold, and were therefore usually found close to more densely populated areas. This location practice, still common today, allowed such businesses to operate more efficiently and economically by servicing larger numbers of people (Christaller [1933] 1966). The specialization seen in this region helps underscore the importance of some retail functions as Christaller showed.

In contrast, the work of moving people and goods shows a different pattern from other occupations. Teamsters and other transport workers (e.g., sailors and toll keepers) were often located away from the major towns (fig. 10.4). For

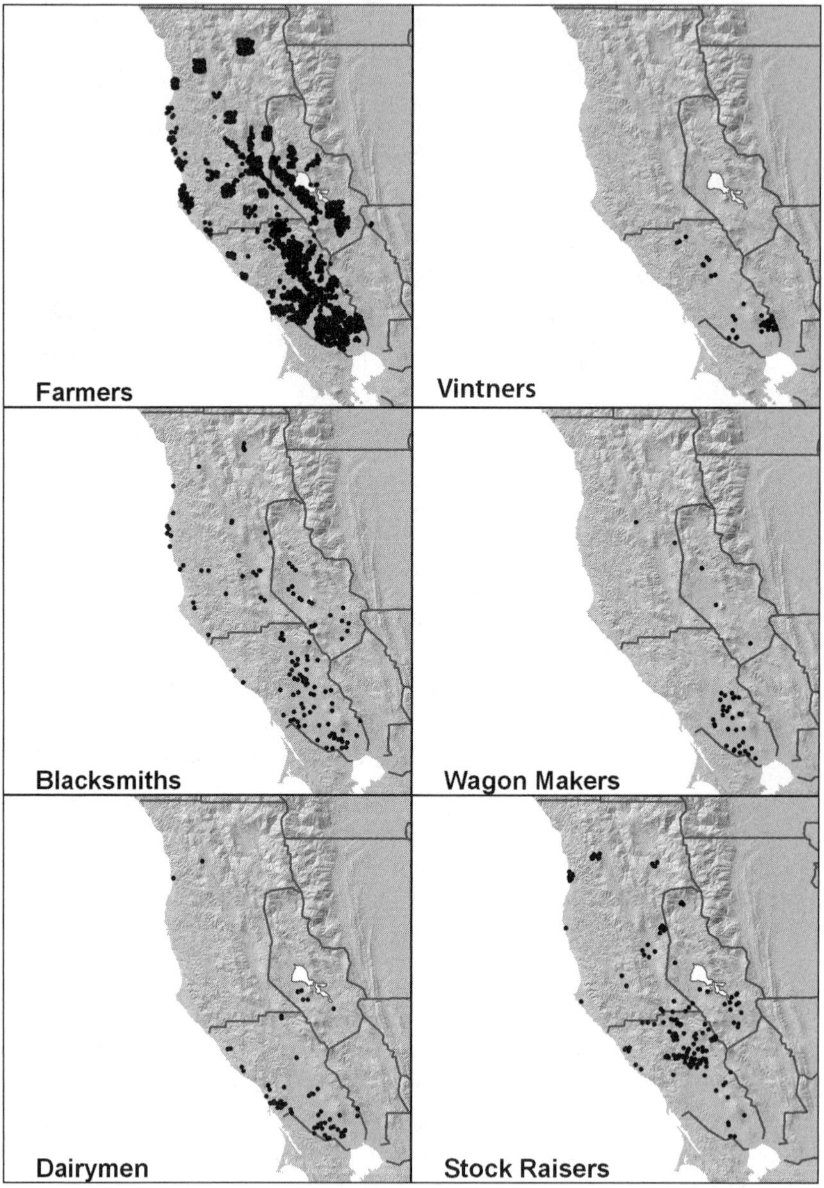

FIGURE 10.3. Selected agricultural and trade occupations (locations of listings). Data source: Paulson 1874; Spencer Larson, BYU Geography.

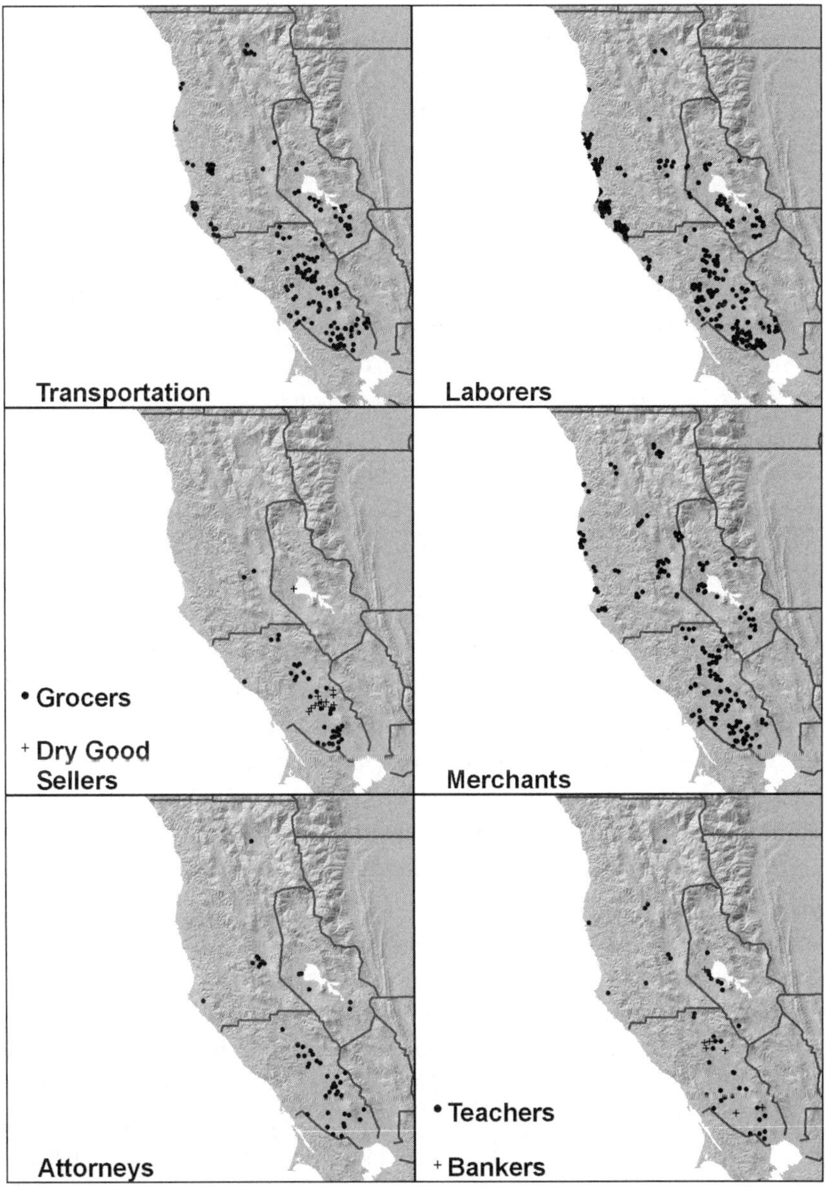

FIGURE 10.4. Laborers and selected service occupations (locations of listings). Data source: Paulson 1874; Spencer Larson, BYU Geography.

example, the large number of teamsters in far-away Round Valley (northeastern Mendocino County) departed sharply from the dearth of listings in Ukiah, the county seat. One reason is that teamsters were in the business of transporting things from a city, port, or railroad depot to remote communities. Many in the transportation sector apparently lived in those villages at the end of a line. Additionally, for the few water transport workers such as the boss lighterman at Navarro Mill, Mendocino County, their places of employ were normally removed from population centers farther inland.

The business of selling goods was very widespread, but self-identified specialized grocers and dry goods proprietors were once again concentrated in Sonoma County (fig. 10.4). In particular, dry goods specialists were located almost exclusively around Santa Rosa, while butchers were more dispersed, roughly paralleling the overall population distribution. These are logical locational choices. Dry goods do not spoil, and business owners could compete and distribute their goods over a large area. Conversely, butchers had a more perishable product that could only be sold within a reasonable transport radius before spoiling. Merchants, as Vance (1970) would emphasize, were spread throughout the three-county region also as they could still profit in smaller markets. The concentration of some retail industries versus the dispersion of wholesale ones and others lends credence to Vance's wholesale model and Christaller's ([1933] 1966) central place paradigm.

Physicians, dentists, teachers, and attorneys had all made their way into the region by 1874 (fig. 10.4). These service occupations tended to concentrate in the towns, although more teachers lived farther out in the country. Once again, the larger population base of Sonoma County allowed it to garner a bigger share of the professionals of the day. The line between the developing core of Sonoma County and the more peripheral status of Lake and Mendocino counties was clearly visible in this occupational area.

Saloons and ministers had a noteworthy spatial configuration. Southern Sonoma County appeared to have more than its fair share of saloons, as did Round Valley in northern Mendocino County. On the other hand, Walker Valley in central Mendocino County seemed to eschew alcohol (no saloons listed with six hotels, though alcohol could have been sold at these hotels). If religiosity can be measured by the number of ministers in residence, then it should be noted here, albeit somewhat facetiously, that Mendocino had nine ministers to the more populated Sonoma's ten. Furthermore, the ratio of ministers to saloons was the lowest in Sonoma County (10 ministers to 76 saloons or .13) and much higher in both Lake and Mendocino counties (.31 and .30 respectively).

TABLE 10.4. Chi-square for changes in employment distribution between 1850–1880 for Lake, Mendocino, and Sonoma counties.

	1850–1852	1852–1860	1860–1870	1870–1880
Lake	Insuff. data	Insuff. data	224.8	32.8
Mendocino	Insuff. data	Insuff. data	231.0	14.8
Sonoma	150.8	257.4	47.7	20.2

(Comparison in relative employee levels of laborers, primary sector workers, manufacturing sector workers, and service sector workers [including transportation]. The expected values were from the first year in each period. All significant at the .001 level.)

The data in *Paulson's Hand-book* show the tremendous diversification in the economy that had occurred just twenty-four years after statehood. To be sure, natural resource extraction, especially farming, still held the preeminent position in terms of total numbers employed. However, the sheer growth in number of occupations, and local concentration of some jobs and dispersion of others, indicates that the retail service function was able to reach farther and farther into the hinterland (or the area that a city or county is connected to economically).

Approach to Settlement Structure Stability

It is remarkable to see just how quickly the economies of these counties "settled" into a clearly implanted urban hierarchy, with fairly stable employment ratios. The progression from 1850 to 1880 is quite startling. In 1850, there were fewer than 700 people in the area of modern Lake, Mendocino, and Sonoma counties. In 1880, there were 6,596 people in Lake County, 12,800 in Mendocino, and 25,926 in Sonoma (table 10.1). In other words, Sonoma County, followed by Mendocino and Lake counties, sped from a wilderness state through the frontier period to a developed farming and services structure in just thirty years. With this population boom, there was a total transformation of the economy, spurred by its original resource-extraction base (R. Walker 2001a. For a Canadian example, see Russwurm and Thakur 1981).

Although only thirty years had passed since statehood, this semiperipheral area had already developed a more stable and recognizable structure. There was both an overall approach to employment sector stability, as evidenced by smaller changes among economic sectors, and a noticeable move toward regional integration of cities and townships by function and size. A chi-square analysis comparing the amount of change between years in the relative distribution of jobs among the main employment sectors shows substantial decline in volatility for all three counties (table 10.4). The chi-square values all register significant differences between years, but the great drop in the three counties indicates that there was much less change between 1870 and 1880 than between earlier years. In terms of function, Sonoma County had emerged as the predominant county in population with the largest service sector. Mendocino County followed, being second in number of inhabitants, while it had a similar proportion of people in the primary sector. However, it had many fewer workers in the service sector. Finally, Lake County had evolved into an agricultural and quicksilver mining center complemented by a small recreation/resort economy centered on the county's many mineral springs.

Many towns in the three counties had also grown to noteworthy size by 1880. Santa Rosa, Petaluma, Healdsburg, and Cloverdale had set themselves firmly into settlement hierarchy within Sonoma County. Ukiah in Mendocino County, and Lakeport in Lake County had developed into the most important cities of their counties, partly by virtue of being county seats. Even though these cities had grown substantially, they still were significantly smaller than San Francisco. Vance's explanation that middle-size cities would be less likely in regions where wholesaling is dominant may be applicable in this case. Future urban growth occurred in the budding cities, while some smaller towns were set to shrink (Vance 1970).

Further evidence of significant development during this thirty-year period was the increasing specialization and differentiation that had developed within the counties. The greater size and diversity of the service sector in Sonoma, when compared with the other counties, suggests that the hierarchy posited by Christaller was forming in the region. Additionally, the overall ratio of occupations to total workers increased between 1860 and 1880 in Lake and Sonoma counties, pointing to more relative specialization each year (table 10.5). Using the sampled totals, Mendocino County showed a small drop between 1870 and 1880 and had the lowest specialization ratio of the three counties, possibly indicating its greater reliance on services provided in Sonoma County. Finally, sub-county tabulations indicate that townships had differentiated themselves economically, with each area showing a distinctive economic structure.

Other counties in the region, outside of the main study area, had also achieved significant economic diversification by the time of L. L. Paulson's directories (Paulson 1874; 1875). Santa Clara County, home to San Jose, had over 48 percent of its listings in the service sector, with over 200 different service occupations (table 10.6). This exceeded Sonoma County, which makes sense, as its population in 1870 was higher by about 6,000. It also had the greatest percentage of workers in manufacturing (21.2 percent). Napa, San Benito,

TABLE 10.5. Specialization in occupations in Lake, Mendocino, and Sonoma counties, 1850–1880.

	FULL COUNT 1850	FULL COUNT 1852	FULL COUNT 1860	FULL DIRECTORY 1874	SAMPLE 1860	SAMPLE 1870	SAMPLE 1880
Lake							
Different occupations				102	20	25	50
Number of people				879	360	155	259
Specialization ratio				0.12	0.06	0.16	0.19
Mendocino							
Different occupations	2			161	18	70	72
Number of people	8			1473	166	517	610
Specialization ratio	0.22			0.11	0.11	0.14	0.12
Sonoma							
Different occupations	32	99	230	292		56	82
Number of people	225	1,250	4,489	3,343		370	503
Specialization ratio	0.14	0.08	0.05	0.09		0.15	0.16

(Derived from Paulson 1874 and U.S. Census data. The specialization ratio is the ratio of different occupations to the total number of people for each year.)

and San Mateo counties were more similar to Lake and Mendocino counties in their proportions of manufacturing and service listings, while Monterey had the distinction of being the most tied to the primary sector (with over 66 percent working in that area). Probably owing to the fact that San Benito was formed in 1874, Monterey's listings included those in San Benito County, which means that Monterey County was even more agricultural at that time than the statistics show. Specialization within each of the economic sectors in these counties, as indicated by the numbers of different occupations (especially in the services area), also attests to the complexity of the settlement system. Thus, by the mid-1870s, the region south of San Francisco to Mendocino County had moved quite far toward being well established in settlement structure and function.

SETTLEMENT DIFFUSION PROCESS

Christaller's central place theory does not provide for instances in which a region is not located on a flat plain or a fairly level surface. However, it still can be a basis for understanding variations in the settled environment (Berry and Pred 1961; Everson and FitzGerald 1969; also Beavon 1977; Preston 1992; Skinner 1977). The settlement diffusion history of Sonoma, Mendocino, and Lake counties, with their rugged environments, shows how unevenly settlement can be imprinted on the landscape. As Davies points out, "The precise

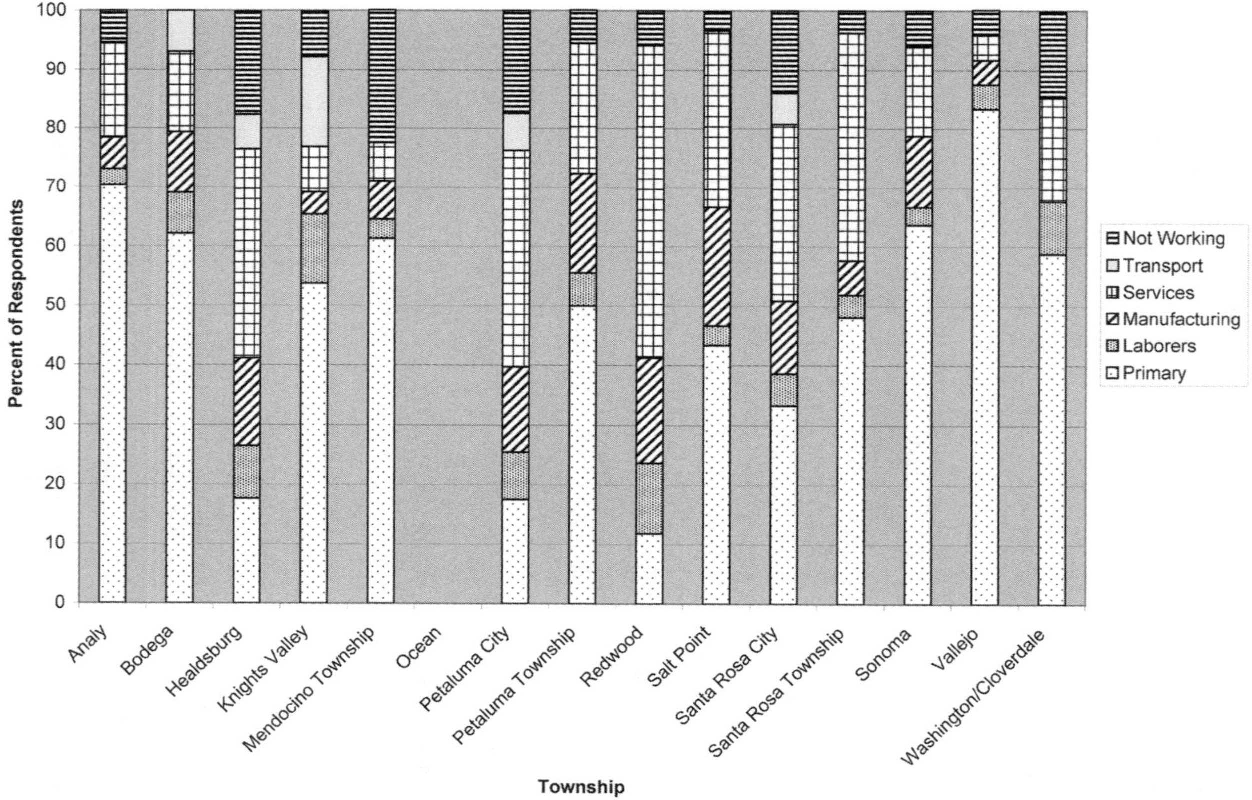

FIGURE 10.5. Relative proportion of workers in different sectors, Sonoma County Townships, 1880 census. Data source: California, Sonoma County 1880.

TABLE 10.6. *Paulson's Hand-book* listings and specialization by major economic sector for Monterey, Napa, San Benito, San Mateo, and Santa Clara counties, 1874–1875 (Paulson 1874, 1875).

	MONTEREY	NAPA	SAN BENITO	SAN MATEO	SANTA CLARA
Primary Sector, without laborers (% of total listings)	66.2	46.3	47.7	36.7	31.4
Laborers %	1.8	7.1	3.8	19.0	6.0
Manufacturing %	9.1	9.7	12.9	12.9	21.2
Services %	18.9	27.2	28.8	21.5	48.4
Transportation %	1.7	2.2	2.1	9.1	3.1
Not specified/Not employed %	2.3	7.2	4.7	.8	11.1
Total Listings	3,321	1,855	906	2,722	3,793
Specialization by economic sector					
Different primary sector occupations	20	9	10	13	17
Laborers (unspecified)	1	1	1	1	1
Different manufacturing occupations	53	43	37	56	97
Different service occupations	103	121	74	107	221
Different transportation occupations	14	11	4	23	21
Total different occupations	191	185	126	200	357

relationships existing in the case study area must be the result of local circumstance" (W. Davies 1967, 77). This admission, however, does not diminish the fact that a hierarchy of settlement sizes and economic diversification diffused very early from San Francisco outward. Far-flung merchants and a lack of middle-size cities at this time help support Vance's wholesaling perspective. Indeed, at a larger scale, distance to San Francisco was a key element in the success of new developments in Sonoma, Mendocino, and Lake counties, as locations closest and most accessible to San Francisco had distinct advantages (particularly in southern Sonoma County).

Here, it is useful to summarize north coast California's experience, as well as that of other areas we have analyzed in the context of the migration and settlement model (fig. 1.2). Up until this point, there are four settlement stages that roughly relate to four of the migration scales in the model. The wilderness stage occurs at the personal and family scales. It begins with immigrants who build isolated homesteads and ends as migration is large enough to create towns. During the frontier stage, which shifts from family to community scale, reliance on farming, mining, and logging continues, but small towns service the needs of the countryside. This is a very volatile time for migration and the settlement diffusion processes. Still, within the community scale, a slowdown in the establishment of new settlements leads to the developed farming and services stage. All areas, especially the core towns, exhibit greater economic diversification and specialization with larger portions of the established and newly arriving population in service and industry, whether dominated by wholesaling or retailing. Finally, the advent of the firmly established settlement stage, within the regional scale of migration analysis, is indicated when the hierarchy of cities and towns is set, and relative employment levels in the different economic sectors exhibit smaller changes over time. New migrants fill in where they find the most opportunity throughout the settlement system. This differentiated structure of towns and farmsteads comes before the relative shift away from natural resource extraction that characterizes the modern urbanization phase in which rural-to-urban migration and overall urban growth are characteristic.

Although I did not measure them directly, I also hypothesize that the population concentration trajectories (or relative distribution of people) would first show dispersion tendencies as settlers rushed to inhabit the countryside. Later, there would be a shift to higher concentration as the urban places in each county grew and the hierarchical system became firmly established. In the firmly established settlement stage, continued migration to urban areas would help reverse the trend of regional population distribution toward concentration. Rural depopulation in the new century and international immigration translate into larger cities and even higher population concentration levels in the twentieth-century urbanization phase.

In this northern California example, I focus primarily on the geographic interface and the spatial perspective of the diffusion process, with less attention to the specific influence of capital (addressed by R. Walker 2001a), immigration, and regional integration. However, the importance of the droves of immigrants, the tremendous capital that was tied to the bounty of the land, and the development of regional connectivity among Sonoma, Mendocino, Lake, and San Francisco counties during the 1850–1880 period must be acknowledged. These functional elements quickly propelled the region through the spatial stages of the model toward an ingrained regional hierarchy of cities and towns by 1880.

The distinct functional/spatial interface of its base geography helped give rise to the region's particular settlement distribution. Settlers favored the main valleys, while most rugged regions found fewer inhabitants unless the lure of redwood encouraged lumber towns. The coastal mountains and ranges surrounding Clear Lake in Lake County acted as barriers that initially hampered development until sufficient transport routes were built to diminish these barriers. Amenable climate and sea access also encouraged the settling of the three counties.

Spatially, the diffusion process had moved from the wilderness stage toward the developed farming and services stage by 1870. During the wilderness period that existed until about 1850 in Sonoma County and into the 1860s in Lake and Mendocino, more settlers found their way into the region, and Sonoma County strengthened its role as the prime service center for the area. Sonoma had a much larger proportion of service workers compared with the other two counties, especially before 1870. In 1870, the frontier stage was already giving way to the developed farming and services phase. Most new towns had been established by that time and the most desirable rural areas had been claimed.

Further development occurred by 1880. Occupational patterns of 1874, from *Paulson's Hand-book*, show how specialized Sonoma had become, as it developed into the dominant county in the region. However, Mendocino and Lake counties were not solely reliant on logging and agriculture but had grown in other economic sectors too, albeit not to the extent of Sonoma. The 1874 analysis also emphasized how particular geographies were associated with the location of various agricultural and commercial enterprises in the counties. Data analysis of the 1880 census shows that further diversification had occurred, but the basic hierarchy of settlements from Sonoma County down to Lake County had experienced less change. The settlement process appears to have moved toward the firmly established stage.

The hierarchical population patterns seen from 1850 to 1880 were subsequently reinforced, as the region moved toward the intense changes of the twentieth-century urbanization stage. Places such as Santa Rosa, Petaluma, and Ukiah continued to grow, but they remained in a class far below San Francisco, which added over 100,000 people to its population from 1880 to 1900. The growth cemented its regional dominance, while Sonoma made its place as the key semiperipheral county, followed by Lake and Mendocino in their even more peripheral roles. Even today, over 130 years after L. L. Paulson printed his directory, this same basic settlement structure from core to periphery is evident in the region.

By way of comparison, the 2010 census shows the tremendous growth of the counties. San Francisco is by far the most densely populated, followed by Sonoma, Mendocino, and Lake. Sonoma, in particular has grown quickly over the past few decades, but it is not close to the 625,000 predicted by Baruth (1960) for the turn of the century, when he called for integrated planning for the rapidly growing greater–San Francisco region. These four counties have therefore continued to experience the same relative population levels and growth patterns that started before 1900. It is noteworthy, however, that Lake

County's population density is actually almost twice that of Mendocino. This stems from the fact that Lake County has been able to take advantage yet again of its attraction as a resort and retirement destination, while Mendocino has been constrained at times by the ups and downs in the lumber industry and its rugged topography (see also Crump 2003). Thus, the rapid diffusion of settlements across these counties that began in the mid-1800s transformed the wilderness into a firmly established hierarchical system that is still very much in place today.

11

Integrating the San Francisco Region from the Pacific Coast to the Sierra Nevada

Little attention has been devoted yet to the growth and development of the primate city of this region: San Francisco. Instead I have attempted to show the importance of the hinterland in providing the early impetus for growth in northern California. But San Francisco grew concurrently with that hinterland. Without the explosion of immigrants who arrived at the mining areas during the first years of the gold rush, San Francisco would not have grown so quickly nor prospered so exceedingly (Soule et al. [1854] 1966, 209–17). Thus, San Francisco's rise was an integral part of the whole and must be considered in concert with the rest of its region (Barth 1975).

Some San Franciscans who had gone to the mines "returned very soon to their old quarters, and found that much greater profits, with far less labor, were to be found in supplying the necessities of the miners, and speculating in real estate" (Soule et al. [1854] 1966, 225). One of the important roles that San Francisco played for its hinterland was as a focal point of trade and commerce. It acted as a great redistribution point that channeled the goods and people of the expansive world of "supply," connected by sea, to the interior mining activity and budding farms of the lands of "demand" that bounded the city on the north, south, and east. In turn, the places of demand in the hinterland became the foci of supply to more distant areas, with San Francisco benefiting from both the comings and the goings. Thus, the fortunes of San Francisco in its early days were inextricably tied to the mines (Brechin 1999, ch. 1). "The discovery of gold guaranteed that San Francisco Bay would witness a degree of urbanization. Transportation facilities of the day and California's geography assured this outcome" (Lotchin 1974, 5). Because the newly developing supply cities of Stockton and Sacramento were on the shallow San Joaquin and Sacramento rivers respectively, San Francisco became the break-in-bulk point where sea-borne shipments were reloaded on smaller boats for the trip inland (Lotchin 1974, 6).

Due to the tremendous demand for mining supplies farther inland, San Francisco (as well as smaller cities such as Sacramento) grew by leaps and bounds.

> During the first part of this year [1849], San Francisco was rapidly increasing the number of housing and population. Every day added sensibly to both. The mines were continuing to yield large returns, most of which was immediately forwarded to San Francisco, in exchange for new supplies.... Meanwhile, others who had been fortunate were returning from the mines with bags of gold dust, to squander in gambling, in drinking and all manner of thoughtless extravagance and dissipation. (Soule et al. [1854] 1966, 223)

The large supply of gold and limited amount of goods and services in San Francisco led to exorbitant prices. For example,

> the price of admission to the pit of the circus was three dollars; while fifty-five dollars was the cost of a private box.... Every mouthful at dinner might be valued at a dime; and to get a hearty meal would cost from two to five dollars.... A small loaf of bread, such as might cost four or six cents in the Atlantic States, brought fifty cents.... It was about as economical to throw away certain soiled articles of clothing and buy new ones, as to get old ones cleaned, when people had to pay ten to twenty dollars for the washing of each dozen articles, large or small. (Soule et al. [1854] 1966, 253)

As San Francisco grew from a chaotic mass of busybodies, traders, gamblers—punctuated by devastating fires—into the most dominant urban center on the Pacific Coast, other counties found their own place in the northern California settlement system. This chapter illustrates the interrelated trajectories of the population growth of counties in the region from 1850 to 2000 and how the core or peripheral location, as well as their geographic characteristics, affected growth trajectories over time. By using geographic changes in space/time as markers at which to seek historical explanations of California's development, a distinctive geographic history is revealed. Thus, this chapter analyzes how the differential geographic structure of the San Francisco city-system, stretching from core cities to the peripheral rural hinterlands (the lands connected to San Francisco economically), is a key factor in explaining the variations of population growth experienced among the region's counties in the state's first six decades. Furthermore, this chapter connects modern California population patterns with the early period, as well as with the wider West, to show how they are interrelated.

A historical cross-section of city-system growth in the context of core through periphery development is something that is not often done at a large scale. But through considering multiple counties and evaluating general population swings by category, a classic pattern can be identified to enable comparison. In northern California, these patterns can show how the whole region functioned and help reveal some of the important interconnections among counties with different relative economic conditions.

POPULATION CHANGE IN THE UNITED STATES AND CALIFORNIA

All population growth and decline in cities and rural areas can be attributed to demographic effects such as migration, morbidity, and fertility. A town's population grows when births exceed deaths (demographic), or when more people move in than move out (migration), or both. Studies regarding the relative influence of natural increase and migration abound. However, it is generally acknowledged that both factors work together in the ongoing change in settlement concentration and dispersion.

Here, I explore the interrelation of population concentration and county growth at a regional level by considering the factors influencing population change. Scholarly explanations give some essential background to understand why population concentrated within groups of counties or dispersed at different time periods. In California and the West, good historical records exist for urban development because of the relatively recent American arrival. Using this rich data set, I develop explanations behind the differential growth patterns experienced by cities in the region.

National-level studies focus on factors that led to population concentration and urbanization in various parts of the country. A sampling of this research includes work on advantages tied to early industrialization and innovation (Pred 1966), the impact of transportation improvements (Borchert 1967), the changing nature of central places as regions grow (Muller 1977), and control of the movement and exchange of goods (Meyer 1980).

Because of difficulties related to inaccurate historical maps and county figures for years prior to 1900, there is less published research that specifically focuses on population concentration at the regional or city-system scale for long time periods. However, some work has investigated how cities and their hinterlands have changed in population concentration trends since 1790, overcoming previous stumbling blocks by using historical digital county boundary files with matching census databases (Earle et al. 1999; Otterstrom 2001; 2003a). This chapter builds upon such previous research by analyzing the reasons

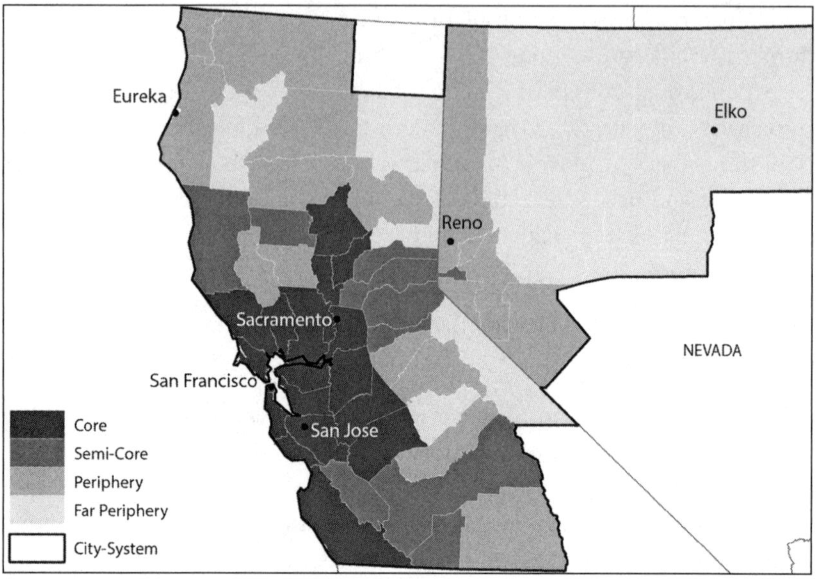

FIGURE 11.1. San Francisco city-system, core to far periphery. Roman Huerta, ThinkSpatial, BYU Geography.

behind population changes and levels of concentration within the San Francisco economic region and neighboring regions over a hundred-year period.

CONCENTRATION AND REGIONAL POPULATION ANALYSIS

I consider the San Francisco region on its own and in context with the other parts of the West. I first analyze population and economic trends in the San Francisco region of counties of the city's primary hinterland and defined as a Basic Trade Area (or what I term a city-system) by Rand McNally's *Commercial Atlas and Marketing Guide* (Rand McNally and Company 1990). It includes all counties of northern California and some in northern Nevada (fig. 11.1). Then, to show how the San Francisco region's counties relate to the greater American West in terms of growth trends, I combine San Francisco with eight additional city-systems: Los Angeles, Portland (Oregon), Salt Lake City, Seattle, Spokane, Phoenix, Denver, and Albuquerque-El Paso (fig. 11.2). These nine regions contain most of the West's area and include relatively older cores, such as Salt Lake City, and newer ones, such as Phoenix. The city-systems, all being in the western United States, also provide examples with similar historical economies and a particularly wide range in population density between central cities and rural counties on the far periphery.

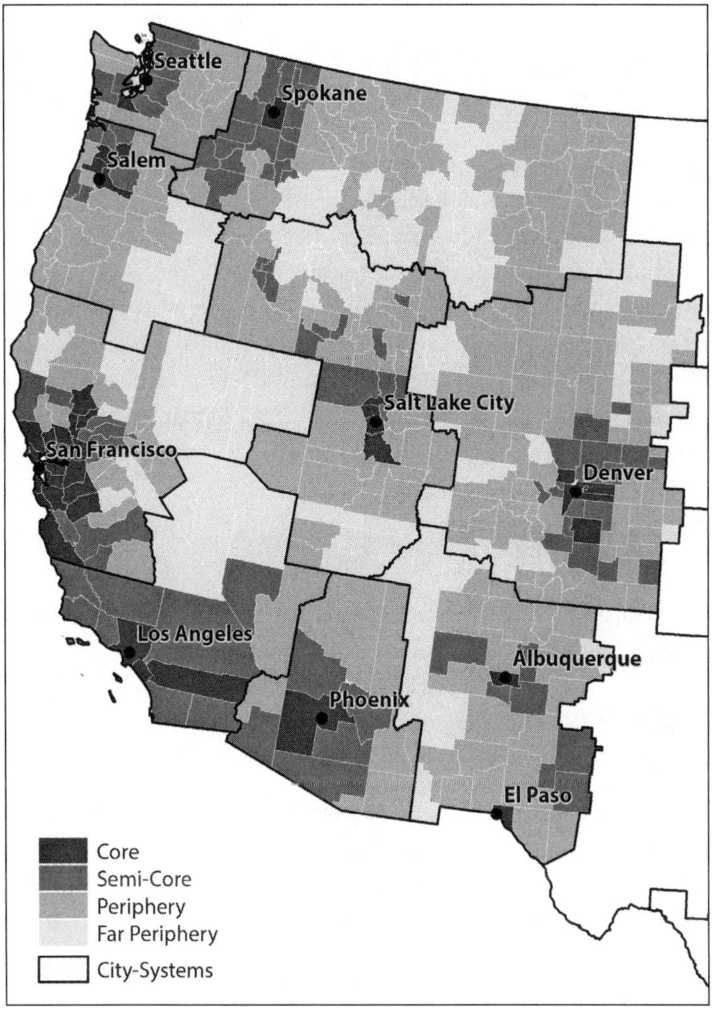

FIGURE 11.2. Western U.S. city-systems, core to far periphery. Roman Huerta, ThinkSpatial, BYU Geography.

To begin, counties were assigned to their respective city-systems from 1850 to 1910. Regional boundaries were projected back to the study period from modern (1990) city-system county clusters to connect each historical region to its modern economic construct. To maintain consistency, these boundaries were kept as firm as possible for each city-system and for the complete study period. Counties that changed shape between years were assigned to the city-system that contained the center point of that county. A classification scheme was then used to assign each county of every region to one of four categories:

core, semi-core, periphery, and far periphery. Four attributes helped differentiate these counties into their respective categories: distance from the core county, relative elevation/ruggedness of topography, and 1990 population density.

Population density is an important measure of human effect on a particular piece of land. Population levels themselves show how many people are in an area but do not give a sense of the relative spread of people unless a total area is included. United States Census researchers realized this early and developed population density categories that matched different types of agricultural and urban structures. An increasing density over time indicates more intense use of the land. If it continues to grow, urbanized areas will be evident. Due to these factors, I used the 1990 population densities, so that the relatively recent settlement intensity can be used to guide the long-term growth trajectories of places within regions.

In the census report section "Progress of the Nation," the five density classes are described as

1. a population of from 2 to 6 to a square mile;
2. a population of from 6 to 18 to a square mile;
3. a population of from 18 to 45 to a square mile;
4. a population of from 45 to 90 to a square mile; and
5. a population of 90 or more to a square mile.

Of different density groups, as they may be called, the first three indicate a predominantly agricultural condition. Speaking broadly, agriculture in the United States is not carried to such a point as to afford employment and support to a population in excess of 45 to a square mile. Consequently, the fourth and fifth groups do not appear except as trade and manufactures arise and the classes rendering personal and professional services are multiplied (Porter et al. 1895, xxx–xxxi).

Using the 1890 census density categories, each county was given a score from 3 to 20 according to its 1990 population density (with 20 being for the densest range of 90-plus people per square mile). The distance factor was weighted less where a maximum of 10 points was possible. The elevation and ruggedness rating also had a maximum of 10 points. After scoring each county, they were divided into four groups for the entire West according to natural breaks, again: core, semi-core, periphery, and far periphery (with the core being those with the highest overall scores). Figure 11.2 shows the distribution of all counties in the West into respective categories. Because of northern California's high population densities and low elevations around San Francisco

TABLE 11.1. Classification of counties in the San Francisco city-system.

COUNTY	CLASS*	SCORE**	COUNTY	CLASS*	SCORE**
Alameda	1	40	Humboldt	3	25
Contra Costa	1	40	Lake	3	25
Marin	1	40	Tehama	3	24
Napa	1	40	Tulare	3	24
Sacramento	1	40	Calaveras	3	23
San Francisco	1	40	Del Norte	3	23
San Joaquin	1	40	Shasta	3	23
San Mateo	1	40	Siskiyou	3	23
Solano	1	40	Tuolumne	3	23
Sonoma	1	40	Douglas, NV	3	21
Yolo	1	40	Madera	3	21
Merced	1	38	Washoe, NV	3	21
Santa Clara	1	38	Lyon, NV	3	20
Santa Cruz	1	38	Mineral, NV	3	20
Stanislaus	1	38	Plumas	3	18
Sutter	1	38	Story, NV	3	18
Butte	1	36	Colusa	3	16
Yuba	1	36	Churchill, NV	4	15
Monterey	1	35	Lassen	4	15
Fresno	2	34	Sierra	4	15
Kings	2	32	Trinity	4	15
Placer	2	30	Alpine	4	11
Glenn	2	29	Elko, NV	4	11
Mendocino	2	29	Humboldt, NV	4	11
Carson City, NV***	2	28	Pershing, NV	4	11
San Benito	2	28	Mariposa	4	10
Amador	2	26	Mono	4	10
El Dorado	2	26	Eureka, NV	4	9
Nevada, CA	2	26	Lander, NV	4	9

*1 = core, 2 = semi-core, 3 = periphery, 4 = far periphery

**Score is the sum of three scores: density, distance, and ruggedness.

***Carson City, NV, has been an independent city since 1969.

there are nineteen core counties (table 11.1). The other western regions have relatively fewer counties in the core category.

For each year from 1850 to 2000, the population density for each county was used to calculate density growth rates for decade periods. These trends were averaged within each county type (core to periphery) to give a baseline for comparison among all types of counties and between individual counties and this average. The average and individual trend lines of density growth rates can then be used to highlight and explicate the general population trends from core to peripheral counties within the city-systems and to explore the reasons why some counties had population growth patterns very unlike the norm.

A related measure, the Hoover Index of Population Concentration (Hoover 1941; Otterstrom 2003a), illustrates the overall changes in relative population distributions for the San Francisco economic region. The index ranges from 0 to 100; higher levels indicate greater population concentration in fewer counties; lower levels show a more dispersed population. As the trend shifts from dispersal to concentration, an important marker in the history of the region is revealed. This point of time highlights the change from growth in both rural and urban counties to population increase more focused in counties with larger cities. This general index can help make sense of the seemingly idiosyncratic population changes at the county level.

The analysis builds upon the exploratory nature of basic population change trajectories by including historical explanations of some of the events that influenced population growth or decline in core or periphery areas and in specific counties that did or did not follow general trends. Population growth is only one characteristic of a county or region, but it gives a consistent indication of its settled condition without having to rely upon other data, which may be incomplete. However, even though raw population numbers are most accessible, the explanation of why they shift over time in a certain area can be a bit of a challenge.

CORE-TO-PERIPHERY GROWTH HISTORY

Growth in the San Francisco city-system has varied greatly between core and periphery since 1850. In all counties, growth was off the chart during the 1850s, as the migration of miners, merchants, and farmers was truly remarkable. Given the small 1850 base population in most counties, in relative terms there was a veritable population explosion in the next decade. This was a time of frontier growth and development. Also, the numbers for that decade are less reliable due to significant boundary changes.

In the 1860s, the core counties still had impressive growth averaging over 6 percent per year. San Francisco County almost tripled its inhabitants from 56,802 to 149,473. The semi-core counties (which include the central Sierra Nevada mining counties) had relatively slower growth than the periphery. The far periphery experienced a reversal during the decade with its three California counties losing population (Mariposa, Sierra, Trinity) as mining and logging growth slowed (fig. 11.3).

Frontier economies enjoyed resurgence in the 1870s, with periphery and far periphery outgrowing core areas. This was a time of extensive growth in rural California and parts of Nevada, with mining, ranching, and farming areas experiencing significant inflow of new settlers. For example, primarily agricultural Tehama County, California, saw over 9.5 percent density growth each year. This did not last long, however, as the peripheries lost population or had flat growth in the ensuing decade of the 1880s.

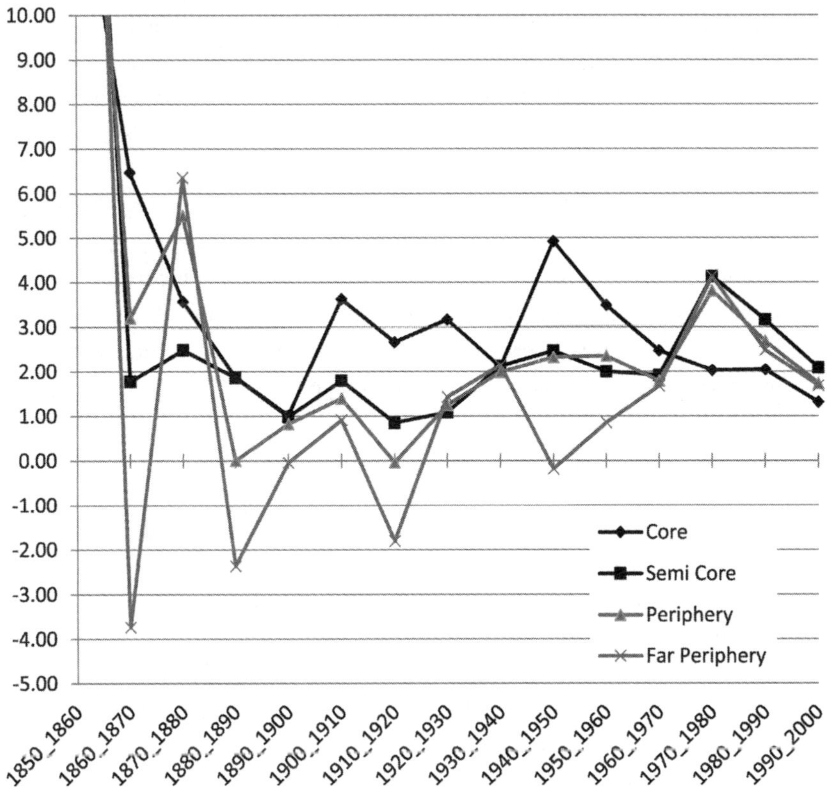

FIGURE 11.3. San Francisco city-system average core-to-periphery population density growth rates.

But the "wild west" was very much alive during the 1870s. In Summit Valley, far northeast Mendocino County, brothers Pierce and Frank Asbill, who were among the early settlers of the valley, built a thriving livestock operation north of Covelo. Legend has it that they hosted the infamous Black Bart in October 1878, a few days after he held up a stage between the towns of Willits and Ukiah. At that dinner, they also fed a preacher, the sheriff, and a beef buyer. Only the Asbill brothers recognized the robber, and they didn't tell, going so far as to put the preacher and Black Bart in the same room for the night (Carranco and Beard 1981, 184–85).

But the Asbills also provide an example of the transformation of the economy at this time. During the 1870s, they and other large ranchers in the area, such as George E. White who also had a colorful history, made large profits on wool, partly owing to high tariffs imposed on imports after the Civil War. An observer noted that this helped the Asbills become "wealthy by mountain standards. And as they grew wealthy, their appearance changed also. In contrast to their long hair and buckskin clothes and moccasins of their early days, they now wore tailor made clothes ordered from San Francisco" (Carranco and Beard 1981, 185–86). These changes in this remote corner of Mendocino County reflect some of the larger shifts of the region from a "wild" frontier to something more "refined" (at least in appearance).

The 1880s and 1890s were a slower time for population growth in much of the city-system, especially on the periphery. On the whole, the far periphery averaged negative population growth in the 1880s (−2.37 percent per year), while the semi-periphery's growth was flat. In contrast, the core and semi-core continued to grow, averaging just under 2 percent per year. In the following decade of the 1890s, each region averaged only between 0 and 1 percent growth per year. It was a time of overall stasis in these parts of California and Nevada. The frontier period ended in California during these years: there was a slowdown in new settlers on the periphery, while the core continued to grow. In fact, the first decade of the twentieth century mirrored this pattern in the country as a whole. The year 1910 marked the end of the frontier in the United States and signaled greater urban growth for decades to come (Otterstrom and Earle 2002).

In California, the decade of 1900 to 1910 brought significant change, especially in core areas. In the core, only Yolo and Sutter counties grew slowly; they were even more agricultural than they are today and closer to some of the semi-core counties in character. Otherwise, the cities were enjoying a boom from an increase in rural-to-urban migration. Sacramento enjoyed over 4 percent annual growth and was representative of this period of intense urbanization.

For the first three decades of the twentieth century, the core grew distinctly faster than the other regions, averaging around 3 percent yearly, while the semi-core, periphery, and far periphery were at least one percentage point lower on average. This urban trend can at least partially be attributed to the rise in manufacturing throughout the regions, primarily in the larger cities. During the 1910s, following the end of the frontier, the periphery and far periphery experienced another decade of negative or flat growth similar to the 1880s. All but two of the far-periphery counties (Lassen and Churchill [Nevada]) lost population in the decade, as the natural resource economies saw an increase in mechanization and reduction in laborers, which resulted in out-migration.

In the midst of this slump on the periphery, between about 1910 and 1912, Hyrum Smith Woolley (great-great-grandfather of the author) attempted to develop a new inland port in Del Norte County, a county in the northwest corner of the state (fig. 11.4). He envisioned that the settlement, named

FIGURE 11.4. Hyrum Smith Woolley and his wife Minerva Rich Woolley, early 1900s (courtesy of Eliza Woolley). Note his curled fingers. He explained this handicap: "The opinion of doctors is that my hands were frosted and re-frosted so many times, both during the time of my contract for the Oregon Shortline Railroad, and also during my stay in Alaska, that the nerves were so affected that the cells became hardened and ossified, finally rendering my hands almost useless. One can well understand that this has been a great handicap for me and it has been very awkward for me to do work of any kind" (H. Woolley 1933, 82–3).

Wooleyport (without the second L), would become the largest city between San Francisco and Portland. The plan was to build a canal from the coast north of present-day Crescent City a few miles inland to Lake Earl and the new port (fig. 11.5). The freshwater would help clean the ships' hulls, and a rail line would connect the city to inland points. Hyrum obtained options on thousands of acres of land surrounding the lake, his son promoted the venture in Grants Pass, Oregon, and Jack London possibly visited the new town at this optimistic time. However, during the process of seeking financial backing, engineers discovered that the lake would have to be dredged to keep it from draining once a sea connection was made. The finding effectively ended the big dreams that Wooleyport would become a "core" city, and the town faded from the map (Woolley C. 1994, 11–14; Woolley H. 1933, 97-103; *Rogue River Courier* 1911). Del Norte County remains a beautiful, peripheral county, still with fewer than 30,000 people.

The Great Depression equalized the region, with average 2 percent population growth rates throughout. With a nod to Steinbeck's *Grapes of Wrath*, many areas of the San Francisco city region were Depression-era destinations, even if

FIGURE 11.5. "Coast of Wooley Port near Crescent City, Cal." ca. 1910. Courtesy of Eliza Woolley.

the overall population grew slowly (Walker 2001b). Then came World War II, a great growth impetus to the core, while the far periphery barely grew at all. This differential growth pattern between core and far periphery was greatest in the 1940s (with the semi-core and periphery hovering in between). Once again, the core benefited from manufacturing growth, some of which was probably tied to the war effort.

The different classes of counties began to converge in the 1950s and 1960s, although the core still had the advantage. Not until the rural revival of the 1970s (Vining and Strauss 1976) did core dominance in growth rates diminish. In the 1970s, the core fell away from the new trend, and in the 1980s and 1990s its growth was still below the other three areas and the differential among the areas narrowed. In the most recent decades, the periphery of the Sierra foothill counties has seen resurgence in growth somewhat reminiscent of the mid-1800s mining boom. But people now come for amenities such as climate, varied topography, and lower population, not the elusive gold in the soil.

Population Concentration

In terms of the Hoover Index of Concentration for the San Francisco region, the geography of the area coupled with evolving growth differences between core and periphery help explain its level (table 11.2). The Hoover Index was highest in 1850 in the midst of the gold rush. A value of 72.5 shows the elevated concentration of people in San Francisco, Sacramento, and the gold-mining counties. Ten years later, the Index had dropped to 66.3, as more of the region

TABLE 11.2. Hoover Index of Population Concentration values for the San Francisco city-system, 1850–2000 (Otterstrom 2003a).

YEAR	HOOVER VALUE	YEAR	HOOVER VALUE
1850	72.46	1930	65.70
1860	66.29	1940	64.26
1870	66.01	1950	66.21
1880	62.10	1960	67.37
1890	61.56	1970	68.19
1900	61.66	1980	65.97
1910	62.26	1990	65.01
1920	65.08	2000	64.28

was settled in places such as Sonoma County, where agriculture was expanding and miners went after their best efforts in the Sierra ended. The population also grew on the periphery, putting more downward pressure on the index.

The 1850 and 1860 numbers were both quite high. (As a reminder, the Hoover Index represents the share of the population that would have to be moved to equalize population density.) But what is interesting about the region is that the Hoover values have oscillated within the 60s ever since 1860. This means that the population of the city-system soon settled into a fairly balanced state, and the small movements that have since occurred show the minor repositioning that has occurred as core or periphery have grown faster or slower than the other. The year 1890 marked the lowest point of the Hoover Index at 61.6, with 1900 very close at 61.7. We can also mark the end of the California frontier with the index, for after 1900 population began to concentrate again.

From 1900 to 1930, the core gained people faster than the periphery, hence the four-point increase in the index to 65.7 by 1930. The index dropped in the 1930s, as growth was more even in its relative rate throughout the region (both in the cities and in the farm- and ranchlands). With the end of World War II, the core grew significantly through urban and suburban expansion, which continued through the 1950s and 1960s as the interstate highway system facilitated housing development on the edges of the core. The second high point of population concentration was reached in 1970 at 68.2. The index then dropped through the year 2000, as the semi-core, periphery, and far periphery all grew relatively faster than the core, illustrating the counterurbanization movement.

COMPARISON WITH THE WEST

In some ways, the San Francisco city-system has similar historical population growth trends to other areas of the West, but in other ways it is quite different. I averaged the decadal growth trends for the regions of Los Angeles, Portland, Seattle, Spokane, Salt Lake City, El Paso-Albuquerque, Denver, and Phoenix,

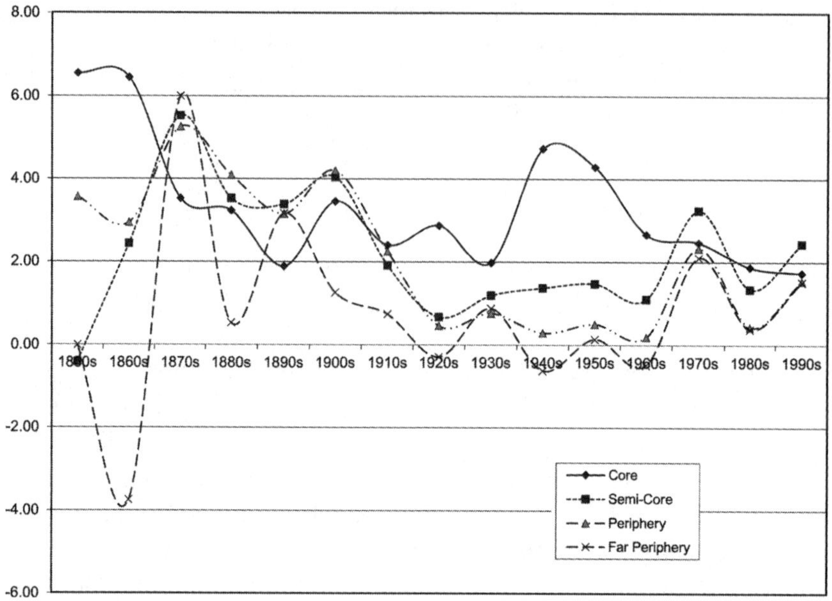

FIGURE 11.6. Western U.S. city-systems average core-to-periphery population density growth rates.

along with San Francisco, to derive a western U.S. view of historical population growth over time (fig. 11.6).

An immediate item of interest is that each city-system has a different mix of core and periphery counties due to varying geographies. For instance, San Francisco has more counties in its core than any of the others, which is due to their proximity to San Francisco, and their relative lower elevation (in the Sacramento Valley and in the San Francisco Bay area). Seattle is a small city-system in area and has only one far-periphery county, while Denver has twenty-five counties in that classification. Spokane and Salt Lake City both have large numbers of periphery counties.

In the West overall, the 1870s was a time of peripheral growth, with the core actually growing the slowest of the four groups. Additionally, from 1880 to 1910 the core lagged behind the semi-core and periphery, but not by a large amount. San Francisco's frontier period ended ten to twenty years earlier than in the West generally. For most of the other regions, 1910 marks the shift to greater core growth relative to the periphery. From 1910 to 1970 the far periphery grew very slowly or contracted (in the 1940s and 1960s), while the core grew the fastest during this period of great urbanization. Things changed in the 1970s in the far periphery and periphery as population growth rates nearly matched the core and semi-core. This is different from San Francisco, where the core dropped out of the growth cluster of the other areas, with population growth about 2 percent lower during that decade.

What continued to be different between San Francisco and the rest of the West for the next twenty years was that growth in the periphery was slower than the core in the West overall, as urbanization in places like Phoenix, Salt Lake City, and Denver continued to be strong. Meanwhile, San Francisco's more established core growth spilled over to adjoining counties in the semi-core and peripheral counties and the core counties did not grow nearly as fast.

INTERRELATED SCALES AND REGIONAL GROWTH

From the individual scale of arrival of a single miner at the busy mining town of Placerville in 1849, to the regional scale of contrasting growth trajectories of core and peripheral counties, the movements of people are interrelated and tied to the stable physical geography of each area and changing economic geographies, locally, regionally, and nationally. At different times, a drop in gold demand changed the fortunes of individual miners in Grass Valley, along with businesses in the town, the county, and in other mining counties in northern California and Nevada. At another time, local conditions such as the availability of good ground for farming in Lake County that resulted in the increase of cultivated land there from 9,026 acres in 1868 to 14,676 acres in 1880 helps explain the population growth of that county from 2,969 in 1870 to 6,596 in 1880 (Palmer [1881] 1995, 86).

During the past century this interconnected growth has continued. For example, the nationwide depression of the 1930s brought in-migration to California and very similar growth trends among core and peripheral areas. The rural rebound or counterurbanization of the 1970s was a national trend that also influenced population movement in California, with resultant growth in the periphery. Furthermore, suburban growth over the last decades of the 1900s and the first decades of the 2000s has spread higher density settlement ever farther north, south, and east of the San Francisco Bay Area, while real estate prices in the core have skyrocketed. Thus, the San Francisco city-system has been able to consolidate and reinforce its interrelations over the decades. San Francisco, as the dominant commercial hub, and the bountiful natural resources in the periphery continue to work together in loose symbiosis.

The migration and settlement model once again helps to illustrate the processes involved in the regional growth and change of northern California (fig. 1.2). From the larger regional scale, the functional growth rate has been driven by economic conditions that extend beyond the region to the national and international spheres. In-migration to the region during good times translated into differential growth depending on the underlying economic substrate of each county. Genealogical connections are important as well, even at this level, as illustrated by family descendancy research. The example of the Winkler family shows that, once established, the majority of children, grandchildren, and great-grandchildren stayed in the region, being tied economically and socially to the home of their recent ancestors.

Thus, this chapter has shown how at the regional level settlement development has continued through all five stages, illustrated through the variable growth between core and periphery. These changes have led to a much more integrated and interrelated regional system than was evident during the wilderness and frontier periods. It has also placed this book's analyses of the smaller scales of community, social network, family, and individual into spatial and historical context.

NORTHERN CALIFORNIA ALTOGETHER, NOW AND THEN

The foothills of the western Sierra jut from the higher snowy slopes to the east. With arthritic finger ridges grabbing at the flat valleys below, they beckon with beguiling promise to people near and far. In the late 1840s, the pull was gold, its undying luster promising a comfortable life in perpetuity for those lucky enough to discover its hiding places. But in the early decades of the twenty-first century, the attraction of the foothills arises more from its natural attributes than its ability to deliver specific economic reward. More trees and a pleasant summer climate encourage ongoing settlement, despite winding roads and steep grades.

Times change, but some crucial elements persist over the generations. Wherever people live, they work and sometimes struggle to produce enough food or sufficient goods and services to exchange for the necessities of life. Far fewer people in the United States are now engaged in agriculture or other primary sector activities, transportation networks and modes have developed and matured, and the nation's population has many more viable locational options for residence. Still, economy of agglomeration and the expanded range of goods and services found in larger concentrations of people have negated any threat of urban obsolescence, although exurban and certain rural locales have continued to see newcomers.

This book weaves the experiences of many early settlers who initially came to California to find prosperity in gold fields, farms, forests, and burgeoning towns into increasingly larger scales to highlight the combined impact of migrants on the growth of settlements from gold camps to the most dominant city of the region, San Francisco. It examines how regional economic and settlement patterns developed differentially in answer to the population influx upon northern California in the late 1800s. The descendants of these early settlers and the newcomers who arrived during the many decades since 1850 have made the state into what it is today: a densely populated, expansive piece of land filled with challenges and innovation.

In the process we have seen a continually evolving push and pull between areas of high population density and rural areas with rapidly growing population densities. These continually reconfiguring patterns of population density have depended on internal demographic change, migration, and economic restructuring. Today there are thousands who live in the inviting foothills because they can work at home, commute using modern transportation methods, or are retired or out of the labor force. Consequently, the United States of the early decades of the 2000s exhibits population patterns that reflect the growing ability of people to choose where they live. Compared with the gold miners of 1849, whose elusive El Dorado was seldom what they hoped, the new El Dorados of the twenty-first century are those places that can offer more and more amenities and opportunities to the argonauts of today.

In the end, northern California is a place of many scales that, seen as a whole, make it the distinct, evolving, and vibrant region that it is. From the individual miner to the migrating family multitudes, from the socially networked Mormons to the founding and survival or decline of hundreds of towns, and to the integration of a cohesive region, the place is a reflection of these interrelated pasts. The family (genealogical) tie from one generation to the next parallel the geographic stage that persists and binds the region together (fig. 1.2). Thus, the destinies of landscapes and the people who created them are interconnected physically and historically through proximate space across the generations.

Acknowledgments

This book has been many years in the making and would never have come to fruition without a great amount of support. Family summer vacations to northern California for eight years in a row set things in motion by opening my eyes to its amazing geographic history. My wife Lori's constant encouragement over these many years has been crucial. She and our children (Nils, Ethan, Meriah, Kristina, Allen, Larissa, Sally, and Samantha [before our youngest, Joran, was born]) were brave and adventurous enough to move with me for a time to a home in the trees near Pollock Pines, California, while I worked on the book.

The College of Family, Home, and Social Sciences and the Department of Geography—Brigham Young University—made it possible for me to take that sabbatical and to continue work on the manuscript since then. Research funds and other support kept the project moving forward. In particular, my college leaders (including Deans David Magleby and Ben Ogles) and department chairs (Professors Richard Jackson, Matthew Shumway, and Ryan Jensen) have been exceptionally supportive and generous. The BYU Office of Research and Creative Activities also funded my work on California's historical migration modeling.

My mother-in-law, the late Jean Thomas, whose family-history research provided excellent material on the Winkler and Greenwood families, was a consistent boon to my work. She is greatly missed. My own parents Jan Otterstrom, Wendy Powell, and Byron Powell, have also continued to encourage me to bring this work to pass.

I am grateful to many others who have added their input to the project including Laurie Weisler (a fantastic editor), Eliza Woolley, BYU Geography's ThinkSpatial student cartographers, Dr. Leonard Hochberg (who gave input on the book's formulation), Dr. Brandon Plewe (who helped improve the cartographic quality of the maps), the late Dr. Carville Earle (my dissertation adviser who taught me how to think as a geographic historian), Brian Bunker (whose computer programming was key to chapter 7), Rex J. Rowley, who suggested I submit my manuscript to the University of Nevada Press, many BYU

Geography Department student secretaries, numerous research assistants and mapmakers (including Nicholle Dods, Candice DeBray, Chris Maderia, Briona Derrick, Joel Warren, Logan Holman, Jonathan Brooks, Emily Pomeroy, Mark Boeckel, and Spencer Larson), Jeff Grathwohl, Dana Henricks and Rebecca Lown for excellent copyediting and cover design work, respectively, and Antone Pierucci and staff of the Lakeport Historic Courthouse Museum.

Finally, Justin Race and Matt Becker of the University of Nevada Press and the anonymous reviewers have made it possible for the book manuscript to actually be published. Many heartfelt thanks are due to each one of you!

References

Abert, J. W. 1999. *Expedition to the Southwest: An 1845 Reconnaissance of Colorado, New Mexico, Texas, and Oklahoma*. Lincoln: University of Nebraska Press.

Adams, J. W., A. B. Kasakoff, and J. Kok. 2002. Migration over the Life Course in XIXth Century Netherlands and the American North: A Comparative Analysis Based on Genealogies and Population Registers. *Annales de Démographie Historique* 2002(2):5–27.

Anderson, W. 1890. *Mineral Springs and Health Resorts of California: With a Complete Chemical Analysis of Every Important Mineral Water in the World*. San Francisco: Bancroft Company.

Bagley, W., ed. 1992. *Frontiersman: Abner Blackburn's Narrative*. Salt Lake City: University of Utah Press.

———. 1999. *Scoundrel's Tale: The Samuel Brannan Papers*. Spokane, WA: Arthur H. Clark.

———. 2010. *So Rugged and Mountainous: Blazing the Oregon and California Trails, 1812–1848*. Norman: University of Oklahoma Press.

Bailey, A. 2005. *Making Population Geography*. London: Hodder Arnold.

Bailey, P. 1943. *Sam Brannan and the California Mormons*. Los Angeles: Westernlore Press.

Bancroft, H. H. 1888. *History of California*. San Francisco: History Company.

Barth, G. 1975. *Instant Cities: Urbanization and the Rise of San Francisco and Denver*. New York: Oxford University Press.

Baruth, K. H. 1960. *The Northern California Metropolis*. Palo Alto, CA: National Press.

Bean, W., and J. J. Rawls. 1983. *California: An Interpretive History*. 4th ed. New York: McGraw Hill.

Beavon, K. S. O. 1977. *Central Place Theory: A Reinterpretation*. New York: Longman.

Beck, W. A., and Y. D. Haase. 1974. *Historical Atlas of California*. Norman: University of Oklahoma Press.

Berry, B. J. L., and A. Pred. 1961. *Central Place Studies: A Bibliography of Theory and Applications*. Philadelphia, PA: Regional Science Research Institute.

Bieder, R. E. 1973. Kinship as a Factor in Migration. *Journal of Marriage and the Family* 35:429–39.

Bigler, D. L., ed. 1990. *The Gold Discovery Journal of Azariah Smith*. Salt Lake City: University of Utah Press.

Billington, R. A., and M. Ridge. 2001. *Western Expansion: A History of the American Frontier*. 6th ed. Albuquerque: University of New Mexico Press.

Blodgett, P. J. 1999. *Land of Golden Dreams: California in the Gold Rush Decade, 1848–1858*. San Marino, CA: Huntington Library Press.

Board of Supervisors of Lake County. [1888] 1988. *A Description of Lake County California, Showing Its Advantages as a Place of Residence*. Lake County, CA: Clear Lake Press. Reprint, Lake County, CA: Lake County Historical Society. Citations refer to the Historical Society edition.

Boeckel, M. A., and S. M. Otterstrom. 2009. From Wilderness to Megalopolis: A Comparative Analysis of County Level Sex Ratios in the United States from 1790 to 1910 Using a Historical GIS. *Social Science Computer Review* 27(3):297–312.

Boggs, W. M. 1912. Reminiscences of Wm. M. Boggs, Son of Governor Lilburn W. Boggs. *Missouri Historical Review* 6:86–90.

Borchert, J. R. 1967. American Metropolitan Evolution. *Geographical Review* 57:301–32.

Bowden, M. J., B. L. LaRose, and B. Mishara. 1971. The Development of Competition between Central Places on the Frontier: Vermont, 1790–1930. *Proceedings of the Association of American Geographers* 3:32–38.

Bradley, W. W. 1918. *Quicksilver Resources of California*. Sacramento: California State Printing Office.

Brands. H. W. 2002. *The Age of Gold: The California Gold Rush and the New American Dream*. New York: Anchor Books.

Brechin, G. 1999. *Imperial San Francisco: Urban Power, Earthly Ruin*. Berkeley: University of California Press.

Brown, L. A. 1981. *Innovation Diffusion: A New Perspective*. London: Methuen.

Bryant, E. 1848. *What I Saw in California*. New York; D. Appleton.

Bullock, R. H. 2009. *Ship Brooklyn Saints: Their Journey and Early Endeavors in California*, (chapter on Winner family). Sandy, UT: R. H. Bullock.

Burchell, R. A. 1987. Opportunity and the Frontier: Wealth-Holding in Twenty-Six Northern Californian Counties, 1848–1880. *Western Historical Quarterly* 18:177–96.

California Historical Landmarks. 1996. Sacramento: California Department of Parks and Recreation, Office of Historic Preservation.

California, Sonoma County. 1880. 1880 U.S. Census, Population Schedule. Washington, D.C.: National Archives and Records Administration.

Carranco, L., and E. Beard. 1981. *Genocide and Vendetta: The Round Valley Wars of Northern California*. Norman: University of Oklahoma Press.

Carter, H. L. [1965] 1982. Jedediah Smith. In *Mountain Men and Fur Traders of the Far West*, ed. L. R. Hafen, 91–108. Lincoln: University of Nebraska Press.

Christaller, W. [1933] 1966. *Central Places in Southern Germany*. Trans. C. Baskin. Inglewood Cliffs, NJ: Prentice Hall.

Clarke, G. 2011. E-mail correspondence received by the author from Gordon Clarke of FamilySearch, June 26, 2011.

Cone, G. C. 1850. Diary, 1849–1850. Original manuscript. Brigham Young University Library Special Collections, Provo, UT.

Conzen, M. P. 1972. State Business Directories in Historical and Geographical Research. *Historical Geography Newsletter* 2:1–8.

Conzen, M. P., and K. N. Conzen. 1979. Geographical Structure in Nineteenth-Century Urban Retailing: Milwaukee, 1836–90. *Journal of Historical Geography* 5:45–66.

Corran, W. H. L. 1887. *Langley's San Francisco Directory for the Year Commencing May, 1887*. San Francisco: Francis, Valentine, and Co.

Cottrell, W. A. 1910. *Report on the Highland Pacific Railroad*. Unpublished report, San Francisco, CA: August 15, 1910. A copy is located at the Lakeport Historic Courthouse Museum, Lakeport, CA.

Courgeau, D. 1985. Interaction between Spatial Mobility, Family and Career Life-Cycle: A French survey. *European Sociological Review* 1(2):139–62.

Cowan, R. O., and W. E. Homer. 1996. *California Saints: A 150-Year Legacy in the Golden State*. Provo, UT: Brigham Young University, Religious Studies Center.

Coy, O. C. 1923. *The Genesis of California Counties*. Sacramento: California State Printing Office.

Crume, R. 2011. Star Search. *Family Tree Magazine* 12(5):42–47.

Crump, J. R. 2003. Finding a Place in the Country: Exurban and Suburban Development in Sonoma County, California. *Environment and Behavior* 35:187–202.

Davies, J. K. 1984. *Mormon Gold*. Salt Lake City, UT: Olympus Publishing.

Davies, J. K., and L. K. Hansen. 2010. *Mormon Gold*. North Salt Lake City, UT: Granite Mountain Publishing.

Davies, W. K. D. 1967. Centrality and the Central Place Hierarchy. *Urban Studies* 4:61–79.

Dewey, R. L. 1986. *Porter Rockwell: A Biography*. New York: Paramount Books.

Dilsaver, L. M. 1982. *From Boom to Bust: Post–Gold Rush Patterns of Adjustment in a California Mining Region*. 2 vols. PhD diss., Department of Geography and Anthropology, Louisiana State University and Agricultural and Mechanical College, Baton Rouge.

———. 1983. Post–Gold Rush Population Changes in a Sierra Nevada Mining Region. *Yearbook of the Association of Pacific Coast Geographers* 45:101–15.

———. 1986. The Development of Agriculture in a Gold Rush Region. *Yearbook of the Association of Pacific Coast Geographers* 48:67–88.

Dunn, L. C. 1997. Ye Also Ought to Retain in Remembrance. In *Speeches (Brigham Young University) 1996–97*, 93–98. Provo, UT: Brigham Young University.

Durham, W. T. 1997. *Volunteer Forty-Niners: Tennesseans and the California Gold Rush*. Nashville, TN: Vanderbilt University Press.

Earle, C., S. M. Otterstrom, and J. Heppen. 1999. *HUSCO–1790-1999* (Historical United States County Boundary Files on CD-ROM). Baton Rouge: Louisiana State University, Geoscience Publications.

Edwards, P. L. 1989. *The Diary of Philip Leget Edwards: The Great Cattle Drive from California to Oregon in 1837*. Fairfield, WA: Ye Galleon Press.

Estergreen, M. M. 1962. *Kit Carson: A Portrait in Courage*. Norman: University of Oklahoma Press.

Everson, J. A., and B. P. FitzGerald. 1969. *Settlement Patterns*. London: Longman.

Family Search. https://FamilySearch.org/, accessed July 2016.

Feltman, E. R. 1993. *California's Lake County: Places and Postal History*. Lake Grove, OR: The Depot.

Fradkin, P. L. 1995. *The Seven States of California: A Natural and Human History*. New York: Henry Holt.

Fraser, J. P. M. 1879. *History of Solano County*. San Francisco: Wood, Alley, and Co.

Gay, T. 1967. *James W. Marshall, the Discoverer of California Gold: A Biography*. Georgetown, CA: Talisman Press.

Geoble, M. 1972. A History of Boating. *Pomo Bulletin* (May).

Goetzmann, W. H. [1966] 2006. *Exploration and Empire: The Explorer and the Scientist in the Winning of the American West*. New York: History Book Club.

Goss, H. R. 1958. *The Life and Death of a Quicksilver Mine*. Los Angeles: Historical Society of Southern California.

Grigg, D. B. 1977. E. G. Ravenstein and the Laws of Migration. *Journal of Historical Geography* 3(1):41–54.

Gudde, E. G. 1949. *California Place Names: A Geographical Dictionary*. Berkeley: University of California Press.

———. 1975. *California Gold Camps*. Berkeley: University of California Press.

Hafen, L. R., ed. 1972. *The Mountain Men and the Fur Trade of the Far West*. 10 vols. Glendale, CA: Arthur H. Clark Company.

Hansen, H. J., and J. T. Miller. 1962. *Wild Oats in Eden: Sonoma County in the 19th Century*. Santa Rosa, CA: Hooper Printing and Lithograph.

Hansen, L. K. 1988. Voyage of the Brooklyn. *Dialogue: A Journal of Mormon Thought* 21(3):47–72.

Hardwick, S. W. 1990. Using City Directories to Teach Geography. *Journal of Geography* 89:266–71.

Hardwick, S. W., and D. G. Holtgrieve. 1996. *Valley for Dreams: Life and Landscape in the Sacramento Valley*. Lanham, MD: Rowman and Littlefield.

Harris, D. E. 1983, 1992. *Redwood Empire Social History Project: California State Census, 1852, Sonoma County*. Sonoma, CA: Sonoma County.

Hart, W. H. 1888. Diaries, 1852–1888. Original manuscript, Brigham Young University Library Special Collections, Provo, UT.

Hayes, D. 2007. *Historical Atlas of California*. Berkeley: University of California Press.

History of Sonoma County. 1926. 2 vols. Chicago: S. J. Clarke.

Hoerder, D. 2002. *Cultures in Contact: World Migrations in the Second Millennium*. Durham, NC: Duke University Press.

Hoerder, D., and L. P. Moch, eds. 1996. *European Migrants: Global and Local Perspectives*. Boston, MA: Northeastern University Press.

Holland, S. D. 1997. *Sierra Saints: A Brief History of the Mormons in Western El Dorado County, 1847–1997*. Placerville, CA: S. D. Holland Publishing.

Holliday, J. S. 1981. *The World Rushed In: The California Gold Rush Experience*. New York: Simon and Schuster.

———. 1999. *Rush for Riches: Gold Fever and the Making of California*. Berkeley: Oakland Museum of California and the University of California Press.

Hong, K. 2009. The Structural Transformation of the Antebellum Red River Valley Settlement Systems in Louisiana. In *Geography, History, and the American Political*

Economy, ed. J. Heppen and S. M. Otterstrom, 65–97. Lanham, MD: Lexington Books.

Hoover, E. 1941. Interstate Redistribution of Population, 1850–1940. *Journal of Economic History* 1:199–205.

Horner, J. M. 1904a. Adventures of a Pioneer. *Improvement Era* 7:510–15, 571–75, 665–72, 767–72, 849–54.

———. 1904b. Looking Back. *Improvement Era* 8:29–35, 112–17.

———. 1906. Voyage of the Ship "Brooklyn." *Improvement Era* 9:794–98, 890–93.

Johnson, P. C., ed. 1964. *The California Missions: A Pictorial History*. Menlo Park, CA: Lane Book.

Johnson, M. E. 1977. A Day at Seigler Hot Springs. *Pomo Bulletin* (May).

Johnson, K., ed. 1996. *Unfortunate Emigrants: Narratives of the Donner Party*. Logan: Utah State University Press.

———. 2009. Personal correspondence with the author, May 29, 2009.

Kelly, C., and D. L. Morgan. 1965. *Old Greenwood: The Story of Caleb Greenwood: Trapper, Pathfinder, and Early Pioneer*. Georgetown, CA: Talisman Press.

King, R., ed. 2007. *Atlas of Human Migration*. Buffalo, NY: Firefly Books.

Kish, L. 1954. Differentiation in Metropolitan Areas. *American Sociological Review* 19:388–98.

Kok, J. 2010. The Family Factor in Migration Decisions. In *Migration History in World History: Multidisciplinary Approaches*, ed. J. Lucassen, L. Lucassen, and P. Manning, 215–50. Leiden, Netherlands: Brill.

Lavender, D. 1954. *Bent's Fort*. Lincoln: University of Nebraska Press.

———. 1972. *California: Land of New Beginnings*. New York: Harper and Row.

Lewis, O. 1949. *Sea Routes to the Gold Fields: The Migration by Water to California in 1849–1852*. New York: Knopf.

Lewis, R., and C. Benson. 1949. *Stories and Legends of Lake County*. Santa Rosa, CA: Press Democrat.

Lotchin, R. W. 1974. *San Francisco 1846–1856: From Hamlet to City*. New York: Oxford University Press.

Macpherson, A. G. 1984. Migration Fields in a Traditional Highland Community, 1350–1850. *Journal of Historical Geography* 10:1–14.

Mann, R. 1982. *After the Gold Rush: Society in Grass Valley and Nevada City, California 1849–1870*. Stanford, CA: Stanford University Press.

Marriott, L. D. 1979. Lilburn W. Boggs: Interaction with Mormons Following Their Expulsion from Missouri. PhD diss., Department of Secondary and Higher Education and Foundations, Brigham Young University, Provo, UT.

Mauldin, H. Mauldin Manuscripts. n.d. Unpublished collection of historical materials concerning the history of Lake County, California. Mauldin was the Lake County Historian for many years. Copies are in the Lakeport Historic Courthouse Museum and Lake County Library, Lakeport, CA.

Meyer, D. R. 1980. A Dynamic Model of the Integration of Frontier Urban Places into the United States System of Cities. *Economic Geography* 56:120–140.

Miller, C. S., and R. S. Hyslop. 1983. *California: The Geography of Diversity*. Palo Alto, CA: Mayfield Publishing.

Miller, D. C. 1978. *Ghost Towns of California*. Boulder, CO: Pruett Publishing.

Moch, L. P. 2003. *Moving Europeans: Migration in Western Europe since 1650*. 2nd ed. Bloomington: Indiana University Press.

Muller, E. K. 1977. Regional Urbanization and the Selective Growth of Towns in North American Regions. *Journal of Historical Geography* 3:21–39.

Munro-Fraser, J. P. 1880a. *History of Marin County, California*. San Francisco: Alley, Bowen, and Company.

———. 1880b. *History of Sonoma County*. San Francisco: Alley, Bowen, and Company.

———. [1879] 1994. *History of Solano County, California*. Fairfield, CA: J. Stevenson Publishers.

Nadeau, R. 1965. *Ghost Towns and Mining Camps of California*. Los Angeles: Ward Ritchie Press.

Nelson, R. 2011. *Enemy of the Saints: The Biography of Lilburn W. Boggs of Missouri*. Baltimore, MD: PublishAmerica.

Nixon, S. 1966. *Redwood Empire*. New York: Dutton.

North, D. C. 1955. Location Theory and Regional Economic Growth. *Journal of Political Economy* 63:243–58.

Nugent, W. T. 1992. *Crossings: The Great Transatlantic Migrations, 1870-1914*. Bloomington, IN: Indiana University Press.

Otterstrom, S. M. 2001. Trends in National and Regional Population Concentration in the United States from 1790 to 1990: From the Frontier to the Urban Transformation. *Social Science Journal* 38(3):393–407.

———. 2003a. Population Concentration in United States City-Systems from 1790 to 2000: Historical Trends and Current Phases. *Tijdschrift voor Economische en Sociale Geografie* 94(4):492–510.

———. 2003b. The Influences of Transportation and Geography on the Settlement and Development of Lake County, California. *Journal of the West* 42(2):34–44.

———. 2007. Rapid Settlement Diffusion: The Development of the Semi-Peripheral Region North of San Francisco, California: 1850–1880. *Historical Geography* 35:241–68.

Otterstrom, S. M., and B. E. Bunker. 2013. Genealogy, Migration, and the Intertwined Geographies of Personal Pasts. *Annals of the Association of American Geographers* 103(3):544–69.

Otterstrom, S. M., and C. Earle. 2002. The Settlement of the United States from 1790 to 1990: Divergent Rates of Growth and the End of the Frontier. *Journal of Interdisciplinary History* 33(1):59–85.

Owens, K. N. 2002. Gold-Rich Saints. In *Riches for All: The California Gold Rush and the World*, ed. Kenneth N. Owens, 29–54. Lincoln: University of Nebraska Press.

———. 2004. *Gold Rush Saints: California Mormons and the Great Rush for Riches*. Spokane, WA: Arthur H. Clark.

Palmer, L. L. 1880. *History of Mendocino County, California*. San Francisco, CA: Alley, Bowen and Company.

———. 1881. *History of Napa and Lake Counties, California*. San Francisco, CA: Slocum, Bowen, and Company.

———. [1881] 1995. *History of Lake County 1881*. Fresno CA: Word Dancer Press.

Paul, R. 1947. *California Gold: The Beginning of Mining in the Far West*. Cambridge, MA: Harvard University Press.

Paulson, L. L. 1874. *L. L. Paulson's Hand-book and Directory of Napa, Lake, Sonoma, and Mendocino Counties*. San Francisco: Francis and Valentine.

———. 1875. *Handbook and Directory of Santa Clara, San Benito, Santa Cruz, Monterey and San Mateo Counties*. San Francisco: L. L. Paulson.

Pierucci, T. 2015. Railroads Were All the Talk, *Lake County Record-Bee*, August 6, 2015, http://www.record-bee.com/lifestyle/20150806/railroads-were-all-the-talk, accessed August 2016.

———. 2016. The Switzerland of America, *Lake County Record-Bee*, June 2, 2016, http://www.record-bee.com/lifestyle/20160602/the-switzerland-of-america, accessed August 2016.

Porter, R., H. Gannett, and W. Hunt. 1895. "Progress of the Nation," in Report on Population of the United States at the Eleventh Census: 1890, Part I, xviii–xxxiv. Washington, D.C.: Department of the Interior Census Office, U.S. Government Printing Office.

Pratt, A. 1990. *The Journals of Addison Pratt*, ed. S. G. Ellsworth. Salt Lake City: University of Utah Press.

Pred, A. R. 1966. *The Spatial Dynamics of U.S. Urban-Industrial Growth, 1800–1914: Interpretive and Theoretical Essays*. Cambridge, MA: MIT Press.

Preston, R. E. 1992. Christaller's Research on the Geography of Administrative Areas. *Progress in Human Geography* 16:523–39.

Rand McNally and Company. 1990. *Rand McNally Commercial Atlas and Marketing Guide*. 121st edition. Chicago: Rand McNally.

Ravenstein, E. G. [1885, 1889] 1976. *The Laws of Migration*. New York: Arno Press.

Rensch, H. E., E. G. Rensch, and M. B. Hoover. 1932. *Historic Spots in California*. Vol. 2: *Valley and Sierra Counties*. Stanford, CA: Stanford University Press.

Richards, R. 1991. *Historic San Francisco: A Concise History and Guide*. San Francisco: Heritage House.

Rogers, J. H. 1891. *Colusa County*. Orland, California. [Publisher not identified], 1891.

Rogerson, P. A. 2010. *Statistical Methods for Geography: A Student's Guide*. 3rd ed. London: Sage.

Rogue River Courier. 1911. Wants Grants Pass to Examine Harbor, June 9, 1911, http://oregonnews.uoregon.edu/lccn/sn96088281/1911-06-09/ed-1/seq-6/, accessed September 2016.

Rohrer, S. S. 2010. *Wandering Souls: Protestant Migrations in America, 1630–1865*. Chapel Hill: University of North Carolina Press.

Roske, R. J. 1968. *Everyman's Eden: A History of California*. New York: Macmillan.

Russwurm, L. H., and B. Thakur. 1981. Hierarchical and Functional Stability and Change in a Strongly Urbanizing Area of Southwestern Ontario, 1871–1971. *Canadian Geographer* 25:149–66.

Salley, H. E. 1977. *History of California Post Offices: 1849–1976*. La Mesa, CA: Postal History Associates.

Sargent, J. L. 1927. *Amador County History*. Jackson, CA: Amador County Federation of Women's Clubs.

Schurer, K. 1991. The Role of Family in the Process of Migration. In *Migrants, Emigrants, and Immigrants: A Social History of Migration*, ed. C. G. Pooley and I. D. Whyte. New York: Routledge.

Scott, R. 1944. *Samuel Brannan and the Golden Fleece: A Biography*. New York: Macmillan.

Sherman, J. W., and B. H. Sherman. 1969. *Ghost Towns of Arizona*. Norman: University of Oklahoma Press.

ShipBrooklyn.com. 2008. Passengers on the Ship Brooklyn of 1846, http://www.shipbrooklyn.com/passengers.html, accessed July 2016.

Simmons, W. S. 1998. Indian Peoples in California. In *Contested Eden: California before the Gold Rush*, ed. R. A. Gutierrez and R. J. Orsi, 48–77. Berkeley: University of California Press.

Simoons, F. J. 1953. Development of Transportation Routes in the Clear Lake Area, *California Historical Society Quarterly* 32(4):363–71.

Skinner, G. W. 1977. Cities and the Hierarchy of Local Systems. In *The City in Late Imperial China*, ed. G. William Skinner, 275–351. Stanford, CA: Stanford University Press.

Smith, D. W., and W. D. Broderson. 1989. *Soil Survey of Lake County, California*. Washington, D.C.: United States Department of Agriculture, Soil Conservation Service.

Soule, F., J. H. Gihon, and J. Nisbet. [1854] 1966. *The Annals of San Francisco*. Palo Alto, CA: Lewis Osborne.

Starr, K. 1973. *Americans and the California Dream, 1850–1910*. New York: Oxford University Press.

Stellman, L. J. 1953. *Sam Brannan, Builder of San Francisco*. New York: Exposition Press.

Stevenson, R. L. [1884] 1972. *The Silverado Squatters*. Ashland, OR: Lewis Osborne.

Stewart, G. R. [1962] 1983. *The California Trail*. Lincoln: University of Nebraska Press.

Stollsteimer, J. F., R. H. Courtney, and L. L. Sammet. 1975. *Regional Efficiency in the Organization of Agricultural Processing Facilities: An Application to Pear Packing in the Lake County Pear District, California*. Giannini Foundation Monograph No. 35. Berkeley: University of California Division of Agricultural Sciences, California Agricultural Experiment Station.

Thomas, J. W. 1988. *Winkler Family History*. Santa Rosa, CA: Winkler Family Association.

———. 2000. Personal communication with the author, from Thomas's family history research (approximate date).

Truckee History. 2013. www.truckee.com/history/, accessed December 2013.

Turner, F. J. 1953. *The Frontier in American History*. New York: Henry Holt.

United States. 1875. *United States Official Postal Guide*. Washington, D.C.: U.S. Government Printing Office.

Vance, J. E., Jr. 1964. *Geography and Urban Evolution in the San Francisco Bay Area*. Berkeley: Institute of Governmental Studies, University of California.

———. 1970. *The Merchant's World: The Geography of Wholesaling*. Englewood Cliffs, NJ: Prentice-Hall.

Vining, D. R., Jr., and A. Strauss. 1976. *A Demonstration That Current De-concentration Trends Are a Clean Break with Past Trends*. Philadelphia, PA: Regional Science Research Institute.

Walker, A. M. [1965] 1982. Joseph R. Walker. In *Mountain Men and Fur Traders of the Far West*, ed. L. Hafen, 291–310. Lincoln: University of Nebraska Press.

Walker, R. A. 2001a. California's Golden Road to Riches: Natural Resources and Regional Capitalism, 1848–1940, *Annals of the Association of American Geographers* 91:167–99.

———. 2001b. Industry Builds the City: The Suburbanization of Manufacturing in the San Francisco Bay Area, 1850–1940. *Journal of Historical Geography* 27(1):36–57.

Wallerstein, I. 2004. *World-Systems Analysis: An Introduction*. Durham, NC: Duke University Press.

Wilson, I. H. 1965. *William Wolfskill 1798–1866: Frontier Trapper to California Ranchero*. Glendale, CA: Arthur H. Clark.

Winkler, J. R. 2000. *Old Hangtown: A History of Placerville, California from 1848 through 1856*. Placerville, CA: JRW Press.

Woolley, C. C. 1994. *Kenneth Gaynes Woolley*. Self-published. Copy in possession of the author.

Woolley, H. S. 1933. *A Short Sketch of My Life*. Unpublished autobiography of Hyrum Smith Woolley. In possession of the author.

Young, S. 1998. *Beautiful Spas and Hot Springs of California*. San Francisco: Chronicle Books.

About the Author

Samuel M. Otterstrom is professor of geography at Brigham Young University (BYU), holding geography degrees from BYU and Louisiana State University. His dissertation examined the settlement geography of the United States, 1790–1990, including a consideration of the closure of the frontier. He is the editor/coeditor of two books on U.S. historical geography and has written many articles about modern and historical population, migration, and settlement geography. Otterstrom is the founding editor of the H-Net Historical Geography academic network. He has taught a variety of regional and systematic geography courses, historical geography of the United States, land-use planning, and statistics, and has directed study-abroad programs in Spain and England, as well as BYU's Washington Seminar program in Washington, D.C.

Otterstrom was born and raised in Spokane, Washington, and has lived in the West most of his life. His wife Lori is from northern California, and they now make Utah their home. They have nine children and two grandchildren. He has also worked as a planner in Okanogan, Washington, and served a two-year mission for the Church of Jesus Christ of Latter-day Saints in New Mexico and El Paso, Texas. He has been a Scoutmaster and Varsity Scout Coach for the Boy Scouts of America. In his free time, he enjoys being with family, waterskiing, hiking, playing tennis, and researching family history.

Index